SpringerBriefs in Applied Sciences and Technology

PoliMI SpringerBriefs

More information about this subseries at http://www.springer.com/series/11159
http://www.polimi.it

Paola Pucci · Giovanni Vecchio

Enabling Mobilities

Planning Tools for People and Their Mobilities

POLITECNICO
MILANO 1863

Paola Pucci
Department of Architecture
and Urban Studies
Politecnico di Milano
Milan, Italy

Giovanni Vecchio
Department of Transportation Engineering
and Logistics, Centre for Sustainable Urban
Development (CEDEUS)
Pontificia Universidad Católica de Chile
Santiago, Chile

ISSN 2191-530X ISSN 2191-5318 (electronic)
SpringerBriefs in Applied Sciences and Technology
ISSN 2282-2577 ISSN 2282-2585 (electronic)
PoliMI SpringerBriefs
ISBN 978-3-030-19580-9 ISBN 978-3-030-19581-6 (eBook)
https://doi.org/10.1007/978-3-030-19581-6

This Springer imprint is published by the registered company Springer Nature Switzerland AG
The registered company address is: Gewerbestrasse 11, 6330 Cham, Switzerland

Preface

This book investigates how established transport planning tools can evolve to understand and plan for the ever-changing contemporary mobilities that influence the opportunities available to individuals. It discusses existing techniques, revised in the light of the growing interest in the social implications of transport planning decisions: these include analytical tools to interpret consolidated and emerging phenomena, as well as operational tools to tackle new and existing mobility demands and needs.

This book then addresses the implications of everyday mobility for individuals and communities. The result of a continuous exchange between the two authors brings together the results of their various research projects. Referring to different objects, practices and settings, the works presented are connected by an underlying interest in spatial mobilities and their implications in planning and policy.

As for her chapters, Paola Pucci acknowledges the Mapping and Urban Data Lab (DAStU, Politecnico di Milano), in particular Fabio Manfredini and Carmelo Di Rosa, as well as Reyhaneh Akhond and Alessio Praticò whose elaborations contributed to the drafting of Chaps. 3 and 5, respectively.

As for his chapters, Giovanni Vecchio acknowledges the funding received from the Centro de Desarrollo Urbano Sustentable (CEDEUS) at the Pontificia Universidad Católica de Chile (CONICYT/FONDAP 15110020). For Chaps. 2 and 6, he thanks the funding and guidance received within the Ph.D. course in Urban Planning, Design and Policy at the Politecnico di Milano. The reflections in Chap. 4 were developed within the PoliVisu research project, financed by the European Union's Horizon 2020 research and innovation programme under grant agreement No. 769608.

The authors also thank Zachary Jones (Politecnico di Milano) for the generous proofreading of the final manuscript.

Milan, Italy   Paola Pucci
Santiago, Chile   Giovanni Vecchio

Contents

Chapter 1
Enabling Mobilities: Reinterpreting Concepts and Tools

Abstract Mobility is fundamental for enhancing the individual possibility to pursue one's own aspirations. Every person needs to be mobile to reach those opportunities she has reason to value, depending on her needs, attitudes and interests. Despite different individual habits, skills and attitudes, mobility has the same enabling role for any individual, an aspect that the prevailing operational approaches to urban mobility issues nonetheless still tend to overlook. Yet a growing stream of theoretical approaches in the field of mobilities seems to be working in the direction of enhancing individuals' ability to move and, consequently, their ability to access the opportunities they have reason to value. In transport planning and policy approaches, these reflections produced some experiments that have not yet been consolidated as ordinary practice. To address the issue, this book intends to investigate how established transport planning tools can evolve to understand and plan for ever changing contemporary mobilities. As an introduction, this chapter discusses the main conceptual and operational approaches that consider mobility to be an enabling condition. The cases presented in the book intend to introduce how to revisit established tools in light of enhancing the enabling role of mobility, as well as how emerging and experimental tools may influence urban mobility planning and policy approaches.

Keywords Mobility · Accessibility · Social exclusion · Transport policy

1.1 Introduction

Everyday urban mobility is an increasingly varied phenomenon, characterised by manifold forms and unprecedented pervasiveness. The combined mobility of people, objects and information in all of their complex relational dynamics leads to new networked forms of economic and social life (Sheller and Urry 2006) that restructure traditional, static understandings in order to build societies around an assumption of high mobility (Urry 2000). Mobility is composed of many characteristics, involving not only the spatial dimension of movement, but also the networks that support

This chapter was co-authored by Giovanni Vecchio and Paola Pucci.

© The Author(s), under exclusive license to Springer Nature Switzerland AG 2019
P. Pucci and G. Vecchio, *Enabling Mobilities*, PoliMI SpringerBriefs,
https://doi.org/10.1007/978-3-030-19581-6_1

mobility and the capacities required to move (Canzler et al. 2008). The multiplicity of features, forms and meanings makes it necessary to speak of plural mobilities (Urry 2007). Mobilities have a fundamental impact on individuals, because the connections, assemblages and practices that mobilities enable are crucial to shaping contemporary social life (Adey and Bissel 2010).

Manifold practices determine material and immaterial consequences for people and societies (Cresswell and Merriman 2011), conveying individual and collective meanings of their own (Jensen 2009), as well as contributing to the achievement of alternative individual aims and societal ends (Cresswell 2006, Chap. 2). Mobility in particular is fundamental for enhancing the individual possibility to pursue one's own aspirations. Every person needs to be mobile to reach the opportunities she has reason to value, depending on her needs, attitudes and interests. Furthermore, mobility has become an increasingly essential requisite for interacting with urban settings and participating in social life; a fact highlighted already in the Sixties by the renowned planner Melvin Webber (1964, p. 147), according to whom "it is interaction, not place, that is the essence of the city and of city life". Despite being different for each person, mobility has the same enabling role for any individual.

The operational approaches to urban mobility issues still tend to overlook the enabling role of mobility, that is, the impact it can have on the opportunities available to each person. On the one hand, mainstream transport planning approaches prove ineffective in fully conveying the social impacts of mobility. They tend to privilege the provision of infrastructures and services, assuring an economically sustainable implementation and an efficient management of their operation (Martens 2006). On the other hand, ongoing social and spatial transformations challenge some traditional assumptions that shape the understanding and planning of mobility. Emerging mobility habits question key principles of the traditional 'utilitarian approach' to transport planning such as the concept of travel as a derived demand or the users' objective of travel cost minimization. The same spatial setting traditionally defined as 'urban' is changing, together with the forms of mobility associated with it. The urban no longer corresponds to a fixed settlement type, rather it conveys a specific form of life (Brenner and Schmid 2015), one no longer identified only with spatial entities comprised within city boundaries (Angelo and Wachsmuth 2015). In such expanded urban settings, traditional infrastructures—also in relation to mobility—operate at new scales, generating fragmentations and inequalities, as well as occasions for experimentation and innovation (Filion and Keil 2017). Similarly, more and more widespread innovative technologies, especially digital networks and mobile devices, provide unprecedented occasions for mobility that nonetheless are unequally available to different populations and territories, raising significant issues of fairness (Docherty et al. 2017).

Planning and policy approaches may thus have to cope with traditional and emerging questions with the aim of enabling mobilities, that is enhancing individuals' ability to move and consequently access the opportunities they value. A growing stream of theoretical and operational approaches, in the field of mobilities, seems to be working in this direction.

1.2 Overcoming the Limits of the Utilitarian Approach to Mobilities

The main evaluative approaches to transport systems tend to overlook the impact that mobility has on the individual ability to reach valued opportunities. Traditional approaches privilege the internal performances of transport systems, pursuing an efficient use of the available resources on the basis of a 'utilitarian approach' (Chu et al. 1992; Litman 2013). Consequently, people are considered mainly according to how their individual features—needs, opportunities and abilities—influence travel behaviours (Dijst et al. 2013), as well as transport volumes, impacting their personal ability to access activities or destinations (Van Wee 2013b). Individual travel behaviours are usually expressed by travel demand, which depends on the utility of the activity to be reached and the costs to reach that destination (Van Wee 2013a). Despite the presence of different elements contributing to individual travel behaviours, the utilitarian approach that emphasises travel demand is the foundation of the four-step model commonly used in transport modelling (Martens 2006).

Since current transport planning and policy approaches are based on demand rather than needs, "current travel patterns are a reflection of the way in which transport resources have been distributed in the past" (Martens 2006, p. 6). Therefore, these solutions are often unable to address the problems of social disadvantage and exclusion deriving from missed mobility opportunities (Van Wee 2011). Meanwhile, other approaches have tried to address the social and ethical concerns related to manifold forms of mobility. Three main clusters can be recognised: ethical approaches related to justice, general approaches focused on accessibility and individual approaches based on people and their practices.

Issues of justice and equity are increasingly discussed in relation to transport planning and policy. A growing literature on the topic tries to define "not whether transportation planning should be based on principles of justice, but on which principles of justice it should be based" (Martens 2017a, p. 7). Despite recognising the social and ethical implications of transport, the approaches dealing with them are still quite heterogeneous. First, there are different political philosophical theories to reference and various concepts—such as justice, equity, fairness—taken as guides (see Pereira et al. 2017 for an extensive review). Then, both procedural and substantive concerns could be significant in relation to the decision-making process and its outcomes. Even focusing simply on the sole substantive dimension implies different aspects defining what a just transportation system would look like. Though most theoretical approaches focus on the accessibility provided by transport systems (Pereira et al. 2017), others consider the distribution of transport resources (Van Wee and Geurs 2011) and the impact of externalities on each person (Gössling 2016). Also, different ethical principles of reference would influence the choice of certain operational approaches, their features and their distributive outcomes over others (Martens and Golub 2012). In this context, a growing debate on 'mobility justice' is discussing "an overarching concept for thinking about how power and inequality inform the governance and control of movement, shaping the patterns of inequal mobility and immobility in the circulation of people, resources and information" (Sheller 2018, p. 14). In this approach, a "holistic theorization of mobility justice" is proposed to push forward debates around transport equity, spatial justice and sustainable cities toward "a more comprehensive combined analysis of complex intersectional aspects of mobility-related inequities" (Sheller 2018). This vision of mobility justice criticises the distributive rule of transport justice that implies an equitable distribution of access to mobility and of the risks or benefits associated with infrastructures, as proposed by some scholars (Pereira et al. 2017). To replace this approach, the focus on justice suggests addressing an alternative mobility future in which justice might pertain to multiple scales, capable of addressing and developing the fundamental bases for a "full theory of mobility justice" (Sheller 2018, p. 18).

Aggregate analytical approaches consider the social implications of mobility by examining the relationship between transport and social exclusion. Mobility-related exclusion is "the process by which people are prevented from participating in the economic, political and social life of the community because of reduced accessibility to opportunities, services and social networks, due in whole or part to insufficient mobility" (Kenyon et al. 2002, pp. 210–211). The various features and forms of mobility-related social exclusion have led to robust theorisations, evaluative approaches and even policy measures (see Lucas 2012 for an extensive review) that nonetheless struggle with some issues. The same phenomenon receives in fact subtly different conceptualisations, depending on the examined settings and issues, such that poverty, disadvantage and social exclusion are considered in relation to transport supplies, minimum levels of mobility or basic degrees of accessibility (Lucas et al. 2016). Moreover, the deriving evaluative approaches generally suffer from an elevated degree of aggregation, both spatially and socially (Preston and Rajé 2007), that overlooks the specificities of individuals and places. Similarly, it is difficult to assess how accessibility contributes or not to well-being (Stanley and Vella-Brodrick 2009). Finally, despite the rich debate on mobility-related social exclusion and the attempts to reduce the problem in countries like the United Kingdom (Social Exclusion Unit 2003), "there appears to be relatively poor take-up of the transport and social exclusion agenda amongst local transport authorities" (Lucas 2012, p. 112).

A heightened sensitivity to individual features may come from mobility practices. Drawing on the 'new mobilities paradigm' that encompasses people, goods and information (Sheller and Urry 2006), this perspective benefits from approaches more attentive to different practices and experiences. Mobility emerges thus as a dynamic and relational phenomenon shaped by personal, social and spatial features that affect individual opportunities. The relationship between different individual characteristics and the spatial settings experienced on a daily basis define differentiated abilities to be mobile, conveyed by the concept of motility (Flamm and Kaufmann 2006; Kaufmann 2002; Kaufmann et al. 2004). Moreover, these abilities can support diverse practices of access and interaction that involve differentiated spatial scales (Cass et al. 2005; Larsen et al. 2006). Any individual has thus a differentiated potential to be 'mobile', which is an essential condition to appropriate opportunities according to individual purposes. The study of mobility practices themselves is quite widespread (Cresswell and Merriman 2011) and also refers to the "systems lying outside 'individuals'" (Sheller and Urry 2016, p. 14), as well as to the relationships constituting movement (Adey 2006). However, these studies tend to overlook "how practices become meaningful only when they are tied to and allow the completion of activities" (Cass and Faulconbridge 2016, p. 4), for example in relation to access to valued opportunities.

A rich academic literature thus conveys how mobility influences individual opportunities, but theoretical approaches still suffer from practical limitations when defining transport planning strategies. In fact, research on mobility-related social exclusion tends to lack operational dimensions and overlooks "the broader implication of a comprehensive transport policy" (Beyazit 2011, p. 130). Similarly, reflections based on principles of justice have tried to consider how to go beyond evaluations that are simply market-centred (Sen 2000), for example by considering individual needs, values and freedoms within Cost-Benefit Analyses (Martens 2006) and in assessments of the long-term sustainability of transport projects (Van Wee 2011). Nonetheless, mainstream transport planning approaches have resisted such social concerns, also due to their weaknesses in terms of feasibility (Beyazit 2011).

A focus on individual opportunities thereby provides a critical lens to examine the prevailing planning and policy tools in relation to urban mobility. Some explorative attempts have considered the analytical and operational implications of approaches focused on the enabling role of mobility in order to overcome the limitations of the traditional tools for evaluating transport planning. These tend to rely on assessments of displacements based on travel time-cost minimization criteria, for example in trip-based transport models. However, increasing attention is being given to activity-based approaches that consider travel time as 'productive time', focusing on the activities related to each trip and conveying the heterogeneity of the preferences of each person (Dong et al. 2006). Despite their relevance, such approaches lack established applications and robust relationships with real-world planning practice.

Some evaluations have considered specific elements of transport systems: for example, services evaluated according to their contribution to the achievement of a specific list of activities (Inoi and Nitta 2005), or infrastructures designed considering resource location-allocation problems (Wismadi et al. 2014). Other estimations have attempted to define the overall individual freedom to be mobile, evaluating the mobility performance available at the metropolitan scale (Purwanto 2003). Additionally, specific thresholds defining minimum standards may constitute a further evaluative criterion in transport planning. This approach may be useful for measuring the travel time, travel distances and travel expenses that the residents of an area have to face for reaching valued destinations (Hananel and Berechman 2016).

It is instead more difficult to design policy approaches to guarantee better access to urban opportunities. In relation to such a focus, Kronlid (2008, p. 22) mentions two steps required "to develop implementable policy proposals: first, as an ideal understanding of mobility 'unconstrained by limitations' of contextual circumstances, and second, one pragmatic list that takes context into account". As for the first step, Urry's (2007, p. 208) suggestion to consider "intermittent travel and co-presence as a capability" is admittedly difficult to operationalize. Discussing a more general approach, mobility policy should not focus "merely on a person's ability to travel through space", but should rather consider "the possibility of a person to translate the resource into something useful" (Martens and Golub 2012, p. 202). This perspective would shift opportunities into activity participation. While participation represents a functioning, accessibility is seen as capturing a person's capabilities and her freedom to pursue what one has reason to value (Martens 2017b; Van Wee 2011).

Some interventions have contributed in this direction by adopting two different attitudes. On the one hand, universalistic approaches aim at increasing the individual freedom to move, as in the case of free bus travel granted to London youth (Goodman et al. 2014), unemployed and elderly users (for example, in Milan) or even to any user of public transport in cities such as Tallinn and Hasselt (Kębłowski 2018). On the other hand, policy initiatives intend to provide specific mobility opportunities addressing specific areas: for example, some French cities have promoted driving courses in deprived neighbourhoods, where the use of the car could dramatically improve access to the job market (Le Breton 2005; Orfeuil 2004). Meanwhile, other European cities instead have promoted travel training courses for public transport, addressing categories of users with special needs (Bührmann 2010).

1.3 Conceptual and Operational Advancements in Transport Planning Decisions

Many conceptual and operational advancements can contribute to conveying the influence of everyday urban mobility on people and their opportunities. Nonetheless, the brief review of growing theoretical approaches and operations highlights the relevance of stressing the enabling role of mobility and accordingly to re-orientate the mainstream approaches to urban mobility issues. To address the issue, this book intends to investigate how established transport planning tools can evolve to understand and plan for everchanging contemporary mobilities. The research collected in the book deals with existing techniques, revised in light of the growing interest in the social implications of transport planning decisions. These involve both analytical tools to interpret consolidated and emerging phenomena, as well as operational tools to tackle new and existing mobility demands and needs.

Moving from an interest in the implications of everyday mobility for individuals and communities, the chapters presented in the book discuss the relevance of some dynamics and the effectiveness of some tools in enhancing the enabling role of mobility in relation to different settings. Chapter 2 focuses on transport systems, discussing evaluations of the accessibility they provide in light of the access to a basic set of valued activities. Accessibility thus becomes a tool that, through mobility, grants the possibility to achieve significant opportunities to the individual. Chapter 3 describes emerging mobility practices—as Long Distance Commuting—and their temporal space variability as an 'epiphenomenon' of wider transformations in the job market and lifestyles as well as in transport supply and in the organisation of urban settlements that have increased, fragmented and diversified individuals' mobility.

Such emerging mobility practices pose interpretative and operational challenges to deal with their emerging needs, times and conditions of using spaces and networks. Moving to spreading technological innovations, Chap. 4 explores the unprecedented opportunities provided by big data in relation to urban mobility and the representation of flows, as well as the relevant but still emerging issues they raise for fair urban planning and policy approaches. Chapter 5, instead, reconsiders the possibility to evaluate a transport infrastructure—here, railway stations—according to their role as both a node and place of everyday mobility practices. Finally, Chap. 6 discusses the implications that the enablement of mobilities may have for policy approaches, and also the possible introduction of the described tools in real-world planning practice. The conclusions (Chap. 7) propose some open directions for research in terms of orienting evaluative and operational approaches in light of the contribution mobility can give to individuals and their opportunities while addressing policymaking processes in the domain of urban and transport issues.

References

Adey P (2006) If mobility is everything then it is nothing: towards a relational politics of (im)mobilities. Mobilities 1(1):75–94. https://doi.org/10.1080/17450100500489080

Adey P, Bissel D (2010) Mobilities, meetings, and futures: an interview with John Urry. Environ Plan D Soc Space 28:1–16. https://doi.org/10.1068/d3709

Angelo H, Wachsmuth D (2015) Urbanizing urban political ecology: a critique of methodological cityism. Int J Urban Reg Res 39(1):16–27. https://doi.org/10.1111/1468-2427.12105

Beyazit E (2011) Evaluating social justice in transport: lessons to be learned from the capability approach. Transp Rev 31(1):117–134. https://doi.org/10.1080/01441647.2010.504900

Brenner N, Schmid C (2015) The epistemology of urban morphology. City 19(2–3):151–182. https://doi.org/10.1080/13604813.2015.1014712

Bührmann S (2010) Guidelines for implementers of travel training for public transport. Brussels

Canzler W, Kaufmann V, Kesselring S (eds) (2008) Tracing mobilities. Ashgate, Farnham

Cass N, Faulconbridge J (2016) Commuting practices: new insights into modal shift from theories of social practice. Transp Policy 45:1–14. https://doi.org/10.1016/j.tranpol.2015.08.002

Cass N, Shove E, Urry J (2005) Social exclusion, mobility and access. Sociol Rev 53(3):539–555. https://doi.org/10.1111/j.1467-954X.2005.00565.x

Chu X, Fielding GJ, Lamar BW (1992) Measuring transit performance using data envelopment analysis. Transp Res Part A Policy Pract 26(3):223–230. https://doi.org/10.1016/0965-8564(92)90033-4

Cresswell T (2006) On the move: mobility in the modern western world. Routledge, London

Cresswell T, Merriman P (eds) (2011) Geographies of mobilities: practices, spaces, subjects. Ashgate, Farnham

Dijst M, Rietveld P, Steg L (2013) Individual needs, opportunities and travel behaviour: a multidisciplinary perspective based on psychology, economics and geography. In: Van Wee B, Annema JA, Banister D (eds) The transport system and transport policy. Elgar, Celtenham, pp 19–50

Docherty I, Marsden G, Anable J (2017) The governance of smart mobility. Transp Res Part A Policy Pract 115:114–125. https://doi.org/10.1016/j.tra.2017.09.012

Dong X, Ben-Akiva ME, Bowman JL, Walker JL (2006) Moving from trip-based to activity-based measures of accessibility. Transp Res Part A Policy Pract 40(2):163–180. https://doi.org/10.1016/J.TRA.2005.05.002

Filion P, Keil R (2017) Contested infrastructures: tension, inequity and innovation in the global suburb. Urban Policy Res 35(1):7–19. https://doi.org/10.1080/08111146.2016.1187122

Flamm M, Kaufmann V (2006) Operationalising the concept of motility: a qualitative study. Mobilities 1(2):167–189. https://doi.org/10.1080/17450100600726563

Goodman A, Jones A, Roberts H, Steinbach R, Green J (2014) "We can all just get on a bus and go": rethinking independent mobility in the context of the universal provision of free bus travel to young Londoners. Mobilities 9(2):275–293. https://doi.org/10.1136/jech-2012-201753.101

Gössling S (2016) Urban transport justice. J Transp Geogr 54:1–9. https://doi.org/10.1016/J.JTRANGEO.2016.05.002

Hananel R, Berechman J (2016) Justice and transportation decision-making: the capabilities approach. Transp Policy 49:78–85. https://doi.org/10.1016/j.tranpol.2016.04.005

Inoi H, Nitta Y (2005) The planning of the community transport from the viewpoint of well-being: applying Amartya Sen's capability approach. Proc East Asia Soc Transp Stud 5:2330–2341

Jensen OB (2009) Flows of meaning, cultures of movements—urban mobility as meaningful everyday life practice. Mobilities 4(1):139–158. https://doi.org/10.1080/17450100802658002

Kaufmann V (2002) Re-thinking mobility. Ashgate, Farnham

Kaufmann V, Bergmann MM, Joye D (2004) Motility: mobility as capital. Int J Urban Reg Res 28(4):745–756. https://doi.org/10.1111/j.0309-1317.2004.00549.x

Kębłowski W (2018 Aug 24) Public transport can be free. Jacobin

Kenyon S, Lyons G, Rafferty J (2002) Transport and social exclusion: Investigating the possibility of promoting inclusion through virtual mobility. J Transp Geogr 10(3):207–219. https://doi.org/10.1016/S0966-6923(02)00012-1

Kronlid D (2008) Mobility as Capability. In: Uteng TP, Cresswell T (eds) Gendered mobilities. Ashgate, Aldershot, pp 15–34

Larsen J, Axhausen KW, Urry J (2006) Geographies of social networks: meetings, travel and communications. Mobilities 1(2):261–283. https://doi.org/10.1080/17450100600726654

Le Breton É (ed) (2005) Bouger pour s'en sortir: mobilité quotidienne et intégration sociale. Armand Colin, Paris

Litman T (2013) Measuring transport system efficiency. Victoria

Lucas K (2012) Transport and social exclusion: where are we now? Transp Policy 20:105–113. https://doi.org/10.1016/j.tranpol.2012.01.013

Lucas K, Mattioli G, Verlinghieri E, Guzman A (2016) Transport poverty and its adverse social consequences. Proc Inst Civ Eng Transp 169(6):353–365. https://doi.org/10.1680/jtran.15.00073

Martens K (2006) Basing transport planning on principles of social justice. Berkeley Plan J 19:1–17

Martens K (2017a) Transport justice: designing fair transportation systems. Routledge, New York, London

Martens K (2017b) Why accessibility measurement is not merely an option, but an absolute necessity. In: Punto N, Hull A (eds) Accessibility tools and their applications. Routledge, New York, London

Martens K, Golub A (2012) A justice-theoretic exploration of accessibility measures. In: Geurs KT, Krizek KJ, Reggiani A (eds) Accessibility analysis and transport planning: challenges for Europe and North America. Elgar, Celtenham, pp 195–210

Orfeuil J-P (ed) (2004) Transports, pauvretés, exclusions: Pouvoir bouger pour s'en sortir. La Tour d'Aigues: Éditions de l'Aube

Pereira RHM, Schwanen T, Banister D (2017) Distributive justice and equity in transportation. Transp Rev 37(2):170–191. https://doi.org/10.1080/01441647.2016.1257660

Preston J, Rajé F (2007) Accessibility, mobility and transport-related social exclusion. J Transp Geogr 15(3):151–160. https://doi.org/10.1016/j.jtrangeo.2006.05.002

Purwanto AJ (2003) Measuring inequality in mobility: a capability perspective. In: 3rd conference on the capability approach. Pavia

Sen AK (2000) The discipline of cost-benefit analysis. J Legal Stud 29(S2):931–952. https://doi.org/10.1086/468100

Sheller M, Urry J (2006) The new mobilities paradigm. Environ Plan A 38(2):207–226. https://doi.org/10.1068/a37268

Sheller M, Urry J (2016) Mobilizing the new mobilities paradigm. Appl Mobil 1(1):10–25. https://doi.org/10.1080/23800127.2016.1151216

Social Exclusion Unit (2003) Report on transport and social exclusion. London

Sheller M (2018) Mobility Justice. The politics of movement in an age of extremes. Verso, London-Brooklyn

Stanley J, Vella-Brodrick D (2009) The usefulness of social exclusion to inform social policy in transport. Transp Policy 16(3):90–96. https://doi.org/10.1016/j.tranpol.2009.02.003

Urry J (2000) Sociology beyond societies: mobilities for the twenty-first century. Routledge, Abingdon

Urry J (2007) Mobilities. Polity Press, Cambridge

Van Wee B (2011) Transport and ethics: ethics and the evaluation of transport policies and projects. Elgar, Celtenham

Van Wee B (2013a) Land use and transport. In: Van Wee B, Annema JA, Banister D (eds) The transport system and transport policy. Elgar, Celtenham, pp 78–100

Van Wee B (2013b) The traffic and transport system and effects on accessibility, the environment and safety: an introduction. In: Van Wee B, Annema JA, Banister D (eds) The transport system and transport policy. Elgar, Celtenham, pp 4–18

Van Wee B, Geurs K (2011) Discussing equity and social exclusion in accessibility evaluations. Eur J Transp Infrastruct Res 11(4):350–367

Webber M (1964) The urban place and the nonplace urban realm. In: Webber M (ed) Explorations into urban structure. Pennsylvania University Press, Philadelphia

Wismadi A, Zuidgeest M, Brussel M, van Maarseveen M (2014) Spatial Preference Modelling for equitable infrastructure provision: an application of Sen's capability approach. J Geogr Syst 16(1):19–48. https://doi.org/10.1007/s10109-013-0185-4

Chapter 2
Accessibility: Enablement by Access to Valued Opportunities

Abstract Accessibility evaluation is an established technical tool within the domain of transport planning, and it may also have specific significance in relation to the social implications for urban mobility. Definable as the ease of reaching goods, services and activities, accessibility can contribute to defining how transport systems enhance opportunities for individuals by granting them the possibility to participate in different activities that they may have reason to value. The chapter intends to investigate accessibility as a significant evaluative tool for enabling mobilities and define the conditions and adjustments required for enhancing such a role. After introducing traditional definitions in the field of transport planning, accessibility is reconceptualised from a social perspective, focusing on the potential access to basic opportunities that are particularly significant to enable individuals. Moreover, the operational implications for drafting real-world evaluations are discussed. An evaluative exercise referring to the setting of Bogotá (Colombia) is presented to consider how its public transport system grants access to a set of relevant opportunities. The exercise is significant for observing accessibility evaluations in practice, as well as to discuss advantages and limitations of such a technical tool from a social perspective.

Keywords Accessibility · Transport · Equity · Social exclusion · Bogotá

2.1 Introduction

Transport systems are a fundamental contributor to mobility. Mobility in fact does not simply correspond to the spatial dimension of movement, but involves also the individual capacities needed to move and the networks that support mobility (Canzler et al. 2008). Such networks are composed of a hardware component (infrastructures) and a software component (services). Mobility in itself is partially valuable for individuals, but it comes to acquire additional value when related to the achievement of opportunities and the completion of activities that each individual may have reason to value (Chap. 1). Nonetheless, for a long time the main approaches of transportation

---

This chapter was authored by Giovanni Vecchio.

© The Author(s), under exclusive license to Springer Nature Switzerland AG 2019   11
P. Pucci and G. Vecchio, *Enabling Mobilities*, PoliMI SpringerBriefs,
https://doi.org/10.1007/978-3-030-19581-6_2

planning have relied mainly on mobility-oriented analyses, which evaluate transport system performances based on the quantity and the quality of physical travel they can accommodate. Only recently a shift towards accessibility-based analyses has started to appear (Litman 2016).

Accessibility is not an unprecedented concept in transport, but it has increasingly come to assume a central role in planning practice. Introduced in the late 1950s by Hansen's seminal work (1959), accessibility may be defined as "the extent to which land-use and transport systems enable (groups of) individuals to reach activities or destinations by means of a (combination of) transport mode(s)" (Geurs and Van Wee 2004, p. 128). Such a definition is a good starting point to highlight how accessibility is not a concept that stands on its own, but is rather a meaningful feature that significantly impacts the everyday life of urban inhabitants. In this sense, accessibility has been receiving increasing attention from a wide range of transport planning approaches concerned with the social dimensions of mobility.

This chapter intends to investigate accessibility as a significant evaluative tool for enabling mobilities, defining the conditions and adjustments required for enhancing such a role. After introducing traditional definitions in the field of transport planning, accessibility is reconceptualised from a social perspective, focusing on the potential access to basic opportunities that are particularly significant to enable individuals. The chapter also discusses the operational implications for drafting real-world evaluations. An evaluative exercise referring to the setting of Bogotá (Colombia) is presented to consider how its public transport system grants access to a set of relevant opportunities. The exercise is significant for observing accessibility evaluations in practice as well as to discuss the advantages and limitations of such a technical tool from a social perspective.

2.2 A Social Take on Accessibility

Accessibility is a consolidated concept in the field of transport planning, but only recently has it become significant for enabling the role of mobilities. The first definition of accessibility, given by Hansen (1959), focused on the ease of reaching goods, services, activities and destinations, all of which can be defined as opportunities. Accessibility is intended to convey the potential for interaction and exchange, as provided by locations, infrastructures and mobility services also coming more recently to include virtual forms of interaction. The increasing popularity of the concept has led to increasingly complex conceptualisations, of which Geurs and Van Wee (2004) provide an exhaustive example.

Geurs and Van Wee's model contains four components of accessibility: land use, transportation, time and individuals. These elements interact with one another in various ways, influencing themselves and contributing differently to the overall available accessibility. The *land-use* component reflects the land-use system, consisting of the opportunities available at each destination (referring to their amount, quality and spatial distribution); the demand for these opportunities, based on where inhabitants live, and the relationship between the supply and demand. The *transportation* component involves the disutility for an individual to cover the distance between an origin and a destination using a specific mode of transport, including the amount of time, the fixed and variable costs and the necessary efforts. The *temporal* component reflects the temporal constraints, both in terms of available opportunities at different times of the day as well as in terms of individual times for participating in activities. Finally, the *individual* component reflects needs, abilities and opportunities, depending on the features of each person. These characteristics influence how a person can access transport modes and spatially distributed opportunities.

In focusing on individuals and their enablement, mobility becomes increasingly relevant as it influences the possibility that each person has to pursue her own aspirations. Any individual, in fact, has a differentiated potential to be 'mobile', an essential condition to appropriate opportunities according to individual purposes (Kaufmann et al. 2004; Kellerman 2012; Larsen et al. 2006; Urry 2007). Mobility allows access to social resources and, more generally, to participate in social life. Different individual mobilities influence and are influenced by diverse and even unequal forms of accessibility, reflecting that, for a society, "the solution of the problem of access consists in the attempt of organizing itself" (Friedman 2003, p. 100). Different economic, age and gender conditions in fact define diverse individual opportunities for mobility, resulting in potential inequalities and forms of exclusion. These factors emerge when "people are prevented from participating in the economic, political and social life of the community because of reduced accessibility to opportunities, services and social networks, due in whole or part to insufficient mobility" (Kenyon et al. 2002, pp. 210–211).

Accessibility can therefore be crucial to contrast social exclusion (Lucas 2012; Preston and Rajé 2007) and consequently enhance quality of life (Stanley and Vella-Brodrick 2009). In fact, accessibility strongly relates to "the capabilities of performing activities at certain locations" (Van Wee 2011, p. 32). Mobility is instrumental in allowing access to opportunities and participation in activities so that accessibility may become one of the main aims of transport planning (Martens 2017b). In a sense, accessibility also directly involves an ethical dimension: for example, a basic form of accessibility to some destinations "could be labelled as a primary social good" (Van Wee and Geurs 2011, p. 356) and its availability for each individual should be guaranteed by specific distributive policies (Martens 2012). However, the contribution of accessibility to individuals' quality of life and freedom of choice is a relevant, yet unexplored dimension (Van Wee 2016).

Within this framework, transport systems constitute the main—but not only—mediator between people and opportunities. Evaluations of accessibility can thus be significant in considering how transport systems contribute to the enablement of individuals by enhancing or impeding their access to valued opportunities. The contribution that transport systems provide in this sense is different, according to the settings and different populations taken into consideration. Infrastructure in fact may be *in*, *of* or *for* a specific place (Addie 2016), allowing the mobility of some groups while at the same time contributing to the immobility of others (Graham and Marvin 2001). Furthermore, transport systems cannot be simply evaluated in terms of deficit, that is, estimating the available infrastructure. First, transport systems are valuable not in themselves, but rather according to their contribution to individuals' access to opportunities. Second, a number of features (regulations, costs, physical design, integration with the surrounding environment) may determine that the same infrastructure or service is differently available to different groups.

2.3   Shaping Socially-Oriented Accessibility Evaluations

To assess how transport systems may support or not individuals' opportunities, along with their ability to pursue their own freedoms and aspirations, devoted evaluative approaches are necessary. From an enablement perspective, the prevailing evaluative approaches do not have a specific focus on how transport systems contribute to individuals' opportunities. Evaluations of accessibility tend to refer mainly to four categories: infrastructures, locations, people and time (Geurs and Wee 2004). However, the main evaluative approaches to accessibility experience limitations when considering the perspective of individuals, especially in transport planning practice. Transport planners in fact usually privilege infrastructure-based evaluations, in which people are taken into consideration mainly according to the travel behaviours they may develop (Dijst et al. 2013) and their aggregate effects on transport volumes (Van Wee 2013). However, an opposite approach that focuses instead on individuals and their enablement may become significant by considering how urban mobility can enhance or impede individual freedoms and aspirations, thanks to accessibility. Therefore, planning should not focus "merely on a person's ability to travel through space" but rather consider "the possibility of a person to translate the resource into something useful" (Martens and Golub 2012, p. 202). However, accessibility evaluations may be shaped in different ways (see Lucas 2012 for a review), according to diverse ethical principles of reference (Lucas et al. 2016) and, consequently, their distributive outcomes may vary (Martens and Golub 2012). Considering the multiple issues of equity at stake when dealing with accessibility evaluations (Van Wee and Geurs 2011) and giving accessibility an operational dimension, it is necessary to examine at least three issues: accessibility to what, with what and for whom.

As for the opportunities to which it is necessary to guarantee accessibility, it is difficult to provide extensive access to every kind of place, service and activities. Due to these reasons, a focus on sufficient or priority opportunities could be adopted, leading to the pursuit of a 'basic accessibility' (Lucas et al. 2016). Basic accessibility is composed of the activities that "are assumed to be necessary to prevent households from social exclusion" (Lucas et al. 2016, p. 482). According to the examined setting, the set of basic activities may vary consistently. A sufficient threshold of access has to be defined according to how it enhances or places obstacles before individual opportunities. Particularly relevant in this sense are the definition of a set of significant activities to be included within basic accessibility as well as the definition of what is a sufficient level of access to them. As for the definition of the set of activities and opportunities included within basic accessibility, it could be developed through participatory processes or a priori informed analyses. The settling of deliberative procedures around the issues of basic accessibility could be significant (Martens 2017a, Chap. 7) and corroborate the chosen activities with interviews to the subjects living in the analyzed areas.

Equally significant is the definition of a sufficiency standard, which establishes a sufficient level of accessibility. According to the setting, different modal alternatives could be considered (e.g. focusing only on public transport, or on active forms of mobility such as walking and cycling). However, two features in particular require clarification: first, whether accessibility has to be estimated according to the quantity of accessible activities or rather according to the benefits deriving from them; second, the evaluation must consider the potential or real participation in these activities. As for the first issue, it seems that a higher quantity of accessible activities (e.g. jobs) could be significant since "a higher level of accessibility will enable a person to derive more benefit from participating in activities. For example, a higher level of accessibility may enable persons to obtain a job that better matches their skills" (Martens 2017a, p. 113). As for the potential or real participation in activities, if we focus on the choices available to any individual, the potential participation – in terms of accessible activities – appears as a suitable criterion, given that "from the perspective of justice it is crucial to measure a person's possibility to engage in a variety of out-of-home activities" (Martens 2017a, p. 115).

Moreover, the attention on individuals and their enablement suggests that personal features also contribute to the possibility to make use of a transport system; these should enter an evaluation of basic accessibility. Nonetheless, relevant data is often available only at aggregate levels and is therefore unable to convey the individual specificities that could also be significant from an operational point of view. In this sense, some features (for example, the socio-economic conditions of the population living in the examined setting) can be relevant, even if referring exclusively to a collective dimension. Local specificities shall then inform the accessibility measures designed to evaluate a given setting.

2.4 Methodology

An evaluative exercise based in Bogotá (Colombia) is presented here to consider how the public transport system of the city grants access to a set of relevant opportunities[1]. Bogotá was chosen because the city has promoted significant public transport infrastructural investments inspired by an explicit social commitment intended to address significant social imbalances. Nonetheless, the benefits for the worst-off areas and populations have been partial. The experience of Bogotá takes place over almost two decades, with its most significant intervention—the TransMilenio bus rapid transit

---

[1]The analysis presented in this Chapter was conducted in September 2016, so more recent changes to Bogotá's transport system have not been considered. Most of the data used for the analysis are derived from Bogotá's 2015 Origin-Destination survey (Alcaldía Mayor de Bogotá 2017).

system—having started its operations in 2000. Bogotá appeared as a significant setting to test the operational approaches proposed here, even if the contribution of this public transport system to the improvement of social inclusion has been partial, as widely investigated (see for example Ardila-Gómez 2004; Bocarejo et al. 2016; Bocarejo and Oviedo 2012; Gilbert 2008, 2015; Hidalgo et al. 2013; Lotero et al. 2014; Vecchio 2017).

To consider how Bogotá's transport system enhances or impedes accessibility to valued opportunities, the analysis considers how the city public transport system allows access to a set of basic activities, assessing what opportunities are available within a given travel time. The study methodology firstly responds to the three components of accessibility previously described (accessibility to what, with what and for whom).

Assuming a basic definition of accessibility, the analysis considered opportunities that are important from preventing the social exclusion of individuals and households. In general, transport systems are "critical to the welfare of the urban poor and a crucial element in any poverty-oriented city development strategy" (World Bank 2002), but in Bogotá this role seems to be specifically related to work opportunities as more than 70% of daily trips in the city are job-related (Bocarejo and Oviedo 2012). Other significant opportunities are more easily available at the neighbourhood level, as in the case of shopping facilities (Nielsen 2004). For these reasons, the analysis focuses on access to formal employment available in Bogotá at the municipal scale.

The analysis refers to the public transport system of the city, composed by the TransMilenio bus rapid transit and by the Sitp ordinary buses (Spanish acronym for Integrated Public Transport System). Alternative modal choices such as cars, motorbikes and bicycles are not considered, since these are not very significant in the setting of Bogotá for low income groups and for trips less than 15 min in duration (Bocarejo et al. 2016). Instead, walking has been considered as it is necessary for reaching public transport stops due to the irrelevance of intermodality in the city (Rodríguez and Targa 2004, p. 596). The exclusive focus on public transport is quite representative of Bogotá given that low and middle classes represent the vast majority of the population. These groups mainly move by public transport, and buses are the only transport choice available for medium and long-distance trips (Bocarejo and Oviedo 2012).

Accessibility is examined in an aggregate way through a location-based evaluation that considers how many job activities are accessible from each area within a given travel time. The territorial unit of reference are the UPZs (Spanish acronym for Local Planning Unit). Even if a certain degree of approximation is required, the available data easily associates each UPZ to inhabitants with mainly homogeneous socio-economic conditions.

Accessibility is analysed by estimating the potential accessibility to jobs provided by public transport, drawing on Hansen's equation

$$A_i = \sum_j a_j f\left(d_{ij}\right)$$

Ai conveys the accessibility of each origin zone *i*, where inhabitants reside; *aj* conveys the attractiveness of destiny zones *j*, according to the number of jobs that can be found. Finally, $f(dij)$ conveys the function of the distance between origin zones *i* and destination zones *j*. The function of the distance is expressed in terms of time distance, a relevant barrier to activity participation especially in Bogotá. Given the difficulty of assuming a constant travel time budget (Mokhtarian and Cao 2004), two different time thresholds for accessible activities are assumed: 30 min, as in the traditional transport engineering literature (Zahavi 1974), and 60 min, a more realistic value based on the longer commuting habits of the city.

In addition, to consider how the transport system contributes to the available accessibility, potential mobility has been examined. It consists of "the quotient of the aerial of Euclidean distance ('as the crow flies') and the travel time on the transport network between that origin and that destination" (Martens 2017a, pp. 154–155). By considering the lowest distance between two points and comparing it with the travel time on the transport system, potential mobility can highlight whether or not the available transport system performs efficiently.

2.5 Opportunities and Accessibility in Bogotá

The public transport system of Bogotá provides differentiated forms of accessibility in a city characterised by structural imbalances. Imbalances refer to the characteristics of the inhabitants, the distribution of the opportunities, and the availability of public transport. First, the 7.5 million inhabitants of the city are mostly low-income and concentrated prevalently in the southern parts of the city, where informal neighbourhoods and mountainous zones figure predominantly (Fig. 2.1). Second, the distribution of opportunities shows further disparities (Fig. 2.1) concentrated mainly in the central area north of the old town, as well as along the main northern and western road corridors (Bocarejo et al. 2016; Bocarejo and Oviedo 2012; Gutiérrez 2011). Third, the public transport system is differently available to the various neighbourhoods of Bogotá (Fig. 2.2). The Sitp ordinary bus routes serve most of the city neighbourhoods, while only half of the inhabitants have walking access to the TransMilenio (Vecchio 2017).

Moving from the current imbalances of Bogotá, the analysis highlights how different areas of the city have diverse possibilities to access valued opportunities. Job opportunities have been considered by assuming that an area has sufficient accessibility if it can reach at least half of the urban job opportunities within a given travel time. The sufficiency standard is here defined as 50% of the available formal job opportunities and considered in relation to two different travel time thresholds. Three categories of areas emerge: areas with sufficient access when travelling for half an hour by public transport; areas with sufficient access when travelling for one hour and areas with insufficient access in both travel time scenarios. A further measure of potential mobility is offered to examine if eventual low levels of accessibility are caused by the poor performances of the public transport system. As a result,

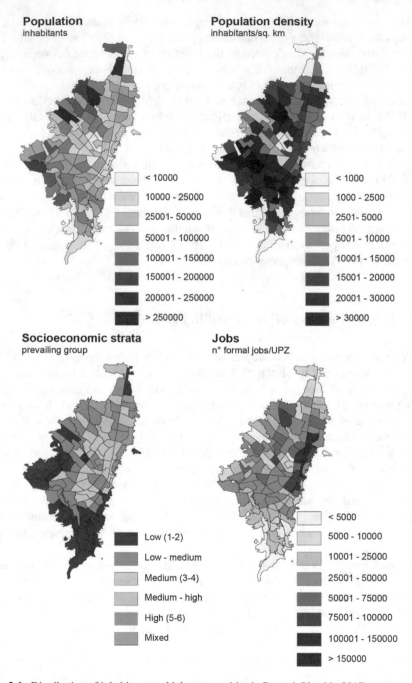

Fig. 2.1 Distribution of inhabitants and job opportunities in Bogotá (Vecchio 2017)

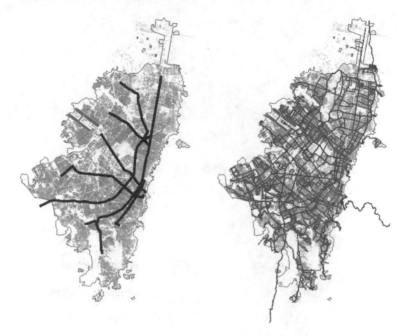

Fig. 2.2 Public transport in Bogotá: TransMilenio (red) and Sitp (blue) (Vecchio 2017)

four different local profiles are defined (Fig. 2.3), distinguishing areas that differ not only because of the accessibility available to them, but also for their socio-economic features and land use patterns.

The consolidated city. The city with sufficient accessibility to jobs within a 30-min travel time distance corresponds to the central nucleus of Bogotá, including its historic centre and immediate proximities. It is the consolidated city which developed immediately before the uncontrolled urban expansion occurred in the second half of the twentieth century. The 11 UPZs that compose it are mainly characterized by a medium population density and by medium-high socioeconomic conditions; these areas are composed of residences, commerce and services. Its distinguishing feature seems to be its central position, close to the most attractive areas where the highest job densities can be found. Therefore, the good availability of the TransMilenio service seems to only add to an already privileged starting condition, where the proximity of the opportunities is simply increased by the BRT system.

The city of TransMilenio. Assuming an accessibility threshold of one hour rather than 30 min, many UPZs (80 out of 112) have a sufficient level of accessibility to the job opportunities of Bogotá. However, only 47 of these also have above-average potential mobility. The area almost perfectly corresponds to the service area of TransMilenio which appears to be the main factor defining the accessibility available to this part of the city. The various UPZs in fact are quite varied if other features are considered. Areas with medium socioeconomic conditions prevail, but

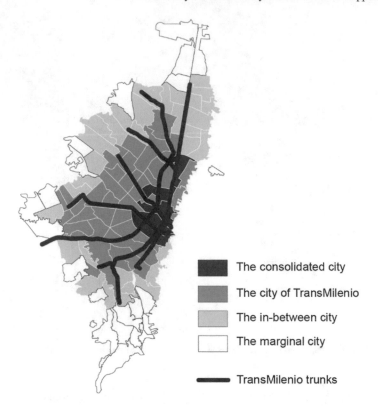

The consolidated city

The city of TransMilenio

The in-between city

The marginal city

TransMilenio trunks

Fig. 2.3 Four local profiles of Bogotá's UPZs, according to accessibility performances

also areas with a prevalence of strata 1 and 2 can be found. Even population density is quite varied, ranging from some of the densest areas of Bogotá to zones with few inhabitants. The prevailing uses include consolidated residential zones, specialised areas (with industrial or commercial locations) as well as developing peripheral areas (both formal and informal).

The in-between city. 33 UPZs are in the peculiar condition of experiencing a below-average potential mobility but which nonetheless maintain sufficient access to job opportunities (when considering the 60 min travel time threshold). Though these areas are in eccentric positions, the category involves different settings: from informal settlements in the southern area to very affluent neighbourhoods in which the highest socioeconomic stratum is dominant. Even population densities and prevailing uses are different, closely recalling the features of the city of TransMilenio. Their peculiar condition may be explained by referring exactly to their eccentric position: from these areas it is not easy to access all parts of the city, but thanks to their position it is possible to reach those zones that offer the most work opportunities. Therefore, they have sufficient access at least when travelling one hour by public transport.

The marginal city. Only 21 areas of Bogotá prove to have insufficient accessibility levels in both the assumed scenarios. These zones seem to be homogeneous, given the features they have in common. First, they occupy a marginal position in the city, being the most peripheral areas. These are also the areas from which it is more difficult to reach the TransMilenio and that tend mainly to rely on other modal alternatives. Given the polarization of Bogotá, where the richest areas are in the northern part of the city and the poorest in the south, it seems that the three northern areas show poor performances because of their development status (they have few built up areas). Instead, the areas in the southern portion of Bogotá are mainly neighbourhoods of informal origins: the prevailing socioeconomic strata are 1 and 2 and most of the UPZs are classified as areas of incomplete urbanization—that is to say, areas of informal origin. They remain unconsolidated and lacking in terms of infrastructures, accessibility, services and public spaces.

The four profiles define general categorizations of Bogotá's UPZs, but nonetheless highlight how various features influence the basic accessibility available in each area. For example, the consolidated city seems to differentiate itself from the other areas due to its location and the dense presence of TransMilenio service. The city of TransMilenio instead mainly relies on public transport to access job opportunities and shows a prevalence of middle-class inhabitants that somehow reflects the fact that most public transport users belong to this group (Lotero et al. 2014). It also shows that the complementary bus routes of Sitp do not much improve accessibility to urban opportunities, an aspect highlighted by other recent evaluations (Guzman et al. 2018). Finally, the marginal city seems to be the expression of its inhabitants' features and needs: most of them are characterized by low socioeconomic conditions and arrive to these areas when fleeing the violence occurring in other regions of Colombia, making informal settlements an initial and precarious answer to basic needs.

2.6 A Place-Based (Re)discussion of Accessibility

The four profiles define general categorizations of Bogotá's UPZs, but nonetheless they highlight how various features influence both the availability to access a set of basic opportunities as well as the resources required for overcoming spatial friction. According to the findings, the profiles may help to redefine basic accessibility considering the emerging peculiarities and, consequently, propose specific priorities for urban mobility planning and policy.

The consolidated city. The city with sufficient accessibility to job opportunities may be reassessed assuming higher sufficiency thresholds. If different thresholds were adopted, it might be necessary to increase local access to valued opportunities. Two lines of action could be relevant in this sense. On the one hand, an improvement of the public transport system (in terms of speed) could be significant. The area already has the highest densities of routes and stops, so that the system's performance could be improved to provide faster connections to other areas offering significant work opportunities. On the other hand, other modal options may be relevant for providing basic access. For example, cycling may be a valuable alternative, thanks to the local presence of cycling paths and the absence of significant differences in altitude. However, the option may be available only to the share of the local population with specific demographic and physical features (excluding for example the elderly or disabled).

The city of TransMilenio. The category includes the majority of Bogotá's UPZs, requiring further specifications to profile its various components and define relevant planning and policy measures. A first distinction may recognize areas that prevailingly originate or attract daily trips: some UPZs in fact have a commercial or industrial focus, while others are mainly residential; a more precise evaluation may assume that their mobility needs are different. Another distinction concerns formal and informal areas, possibly referring to two elements: on the one hand, the presence of different socio-economic groups of inhabitants (and, consequently, different resources available to them); on the other hand, different built environments allow the presence of different transport services so that areas of informal origin may not allow services other than ordinary buses. The reliance of these areas on public transport suggests that a greater integration of the public transport system is needed, focusing on route planning, timetables and fare policy. The first two actions may improve the performance of the service, while fares may ease access to public transport for a wider share of the population. Improving the functioning of the bus network may increase the accessibility available to a wider range of areas, possibly improving basic accessibility even when considering lower time thresholds or assuming higher sufficiency thresholds for the accessible activities.

The in-between city. The eccentric areas that have enough access but experience low potential mobility may call for specific interventions referring to transport. These would be necessary for improving the performances of the current public transport system or in case the sufficiency threshold for basic accessibility were to be increased. Otherwise, the current below-average performances of the transport system do not hinder access to basic opportunities and does not call for specific interventions. The in-between city shows many similarities with the city of TransMilenio, in terms of land use and socioeconomic features. Therefore, similar interventions may be proposed for improving the opportunities for moving in the city, for example by enhancing the integration of the existing public transport routes.

The marginal city. The peripheral settlements characterized by low levels of accessibility share many socio-economic characteristics due to their largely informal origins. Moreover, their lack of sufficient accessibility and the low level of potential mobility call for prioritizing transport interventions in these areas. The impervious features of these areas require adapting tools to evaluate basic accessibility. While larger time thresholds may be more realistic when examining these areas and considering their distance from the central areas, more precise evaluations focused on the various neighbourhoods composing each area could be relevant in considering the longer travel times caused by different altitudes. In any case, the lack of sufficient access marks these as priority areas for intervention. Even by sharpening the evaluative tools, these areas pose difficult conditions for the establishment of regular, reliable public transport services. Yet public efforts still privilege the construction of new infrastructures or the provision of new bus routes that nonetheless have difficulties in meeting local mobility needs. Another option, already attempted by municipal institutions and only partially effective, consists of providing the facilities directly at the neighbourhood scale. Alternative forms of service provision could prove to be significant in these settings, for example by involving the inhabitants in the definition of their mobility needs (and the consequent public transport routes) and even in the coproduction of the needed services (Vecchio 2018a, 2018b).

2.7 Conclusions

Accessibility evaluations emerge as a suitable tool to understand how transport systems can enhance access to valued opportunities and contribute to the enablement of individuals. This approach focuses on the ability to access opportunities and the role of mobility as a significant feature in doing so by using an established technical tool to operationalize the interest for individual opportunities in relation to urban mobility. By aiming at comparing different states and how they benefit or hurt individuals, accessibility allows access to transport systems, mobility services and other modal alternatives according to their contribution to the individual freedom of choice over alternative lives. On the contrary, other indicators that dominate current transport planning—for example, congestion or speed—are relevant only according to how they favour or obscure accessibility (Martens 2017a, Chap. 10).

According to the setting taken into account, the proposed approach requires defining which activities are relevant, which modal choices should be considered and which constraints (for example, in terms of travel time or cost) allow or deny individual access. The benefits for accessibility evaluations are twofold: they first provide a basic form of accessibility that privileges some specific activities that are particularly significant in the perspective of individual opportunities, highlighting priority areas and subjects for intervention (aiming at providing anyone with enough access to those activities that are considered as necessary; Pucci et al. 2019 provide an example in this sense). Second, this focus helps in shaping evaluations that better convey the social impact of accessibility on individuals and their abilities, referring to participation in those activities considered fundamental to take part in society. However, accessibility is prone to some significant limitations: for example, there is a high degree of spatial and social aggregation that overlooks significant local and individual specificities (Preston and Rajé 2007) while critical subjective elements (Cheng and Chen 2015) that define different perceptions of accessibility and valued opportunities are missing.

Accessibility can anyway provide an aggregate look at how transport systems work in relation to the access they provide to certain activities. Such a scale does not allow for very detailed analyses, rather it can highlight areas that are more critical in terms of access to significant urban opportunities. In this sense, basic accessibility emerges as a first tool to direct further, more specific evaluative and operational efforts. Moreover, further specifications are necessary to make it usable within different settings. These refer not only to some evaluative criteria to define when people have enough access to the opportunities of the urban settings they live in, but also to the procedures to establish such thresholds – be them participatory processes including the citizens or simply technical debates involving technicians and policymakers. In addition, different planning and policy measures may emerge as significant options to enhance access in the areas most in need. These may involve the supply of mobility infrastructures and services, even using new forms of provision, the (re)location of needed facilities or even new ways of better coordinating individual needs and mobility offers, involving a governance dimension as well. In any case, accessibility provides a useful tool to consider how urban mobility may enable individuals by making opportunities available to them. This advancement is accomplished by offering a more feasible approach in terms of evaluation, consequent operational directions and contributing to the planning of enabling mobilities.

References

Addie J-PD (2016) Theorising suburban infrastructure: a framework for critical and comparative analysis. Trans Inst Br Geogr 41(3):273–285. https://doi.org/10.1111/tran.12121

Alcaldía Mayor de Bogotá (2017) Encuesta de movilidad 2015. Bogotá

Ardila-Gómez A (2004) Transit planning in Curitiba and Bogotá. Roles in interaction, risk, and change. Ph.D. thesis, Massachusetts Institute of Technology

Bocarejo JP, Escobar D, Oviedo D, Galarza D (2016) Accessibility analysis of the integrated transit system of Bogotá. Int J Sustain Transp 10(4):308–320. https://doi.org/10.1080/15568318.2014.926435

Bocarejo JP, Oviedo D (2012) Transport accessibility and social inequities: a tool for identification of mobility needs and evaluation of transport investments. J Transp Geogr 24:142–154. https://doi.org/10.1016/j.jtrangeo.2011.12.004

Canzler W, Kaufmann V, Kesselring S (eds) (2008) Tracing mobilities. Ashgate, Farnham

Cheng Y-H, Chen S-Y (2015) Perceived accessibility, mobility, and connectivity of public transportation systems. Transp Res Part A Policy Pract 77:386–403. https://doi.org/10.1016/j.tra.2015.05.003

Dijst M, Rietveld P, Steg L (2013) Individual needs, opportunities and travel behaviour: a multidisciplinary perspective based on psychology, economics and geography. In: Van Wee B, Annema JA, Banister D (eds) The transport system and transport policy. Elgar, Celtenham, pp 19–50

Friedman Y (2003) Utopie realizzabili. Quodlibet, Macerata

Geurs KT, Van Wee B (2004) Accessibility evaluation of land-use and transport strategies: review and research directions. J Transp Geogr 12:127–140. https://doi.org/10.1016/j.jtrangeo.2003.10.005

Gilbert A (2008) Bus rapid transit: is TransMilenio a miracle cure? Transp Rev 28(4):439–467. https://doi.org/10.1080/01441640701785733

Gilbert A (2015) Urban governance in the South: how did Bogotá lose its shine? Urban Stud 52(4):665–684. https://doi.org/10.1177/0042098014527484

Graham S, Marvin S (2001) Splintering urbanism. Networked infrastructures, technological mobilities and the urban condition. Routledge, London

Gutiérrez D (2011) Determinantes de la localización del empleo urbano en Bogotá, Colombia. Rev Econ Rosario 14(1):61–98

Guzman L, Oviedo D, Cardona R (2018) Accessibility changes: analysis of the integrated public transport system of Bogotá. Sustainability 10(11):3958. https://doi.org/10.3390/su10113958

Hansen WG (1959) How accessibility shapes land use. J Am Inst Plan 25(2):73–76. https://doi.org/10.1080/01944365908978307

Hidalgo D, Pereira L, Estupiñán N, Jiménez PL (2013) TransMilenio BRT system in Bogotá, high performance and positive impact—main results of an ex-post evaluation. Res Transp Econ 39(1):133–138. https://doi.org/10.1016/j.retrec.2012.06.005

Kaufmann V, Bergmann MM, Joye D (2004) Motility: mobility as capital. Int J Urban Reg Res 28(4):745–756. https://doi.org/10.1111/j.0309-1317.2004.00549.x

Kellerman A (2012) Potential mobilities. Mobilities 7(1):171–183. https://doi.org/10.1080/17450101.2012.631817

Kenyon S, Lyons G, Rafferty J (2002) Transport and social exclusion: Investigating the possibility of promoting inclusion through virtual mobility. J Transp Geogr 10(3):207–219. https://doi.org/10.1016/S0966-6923(02)00012-1

Larsen J, Axhausen KW, Urry J (2006) Geographies of social networks: meetings, travel and communications. Mobilities 1(2):261–283. https://doi.org/10.1080/17450100600726654

Litman T (2016) Evaluating accessibility for transportation planning. measuring people's ability to reach desired goods and activities. Victoria

Lotero L, Cadillo A, Hurtado R, Gómez-Gardénes J (2014) Socioeconomic differences in urban mobility. In: Garas A (ed) Interconnected networks. Springer, Berlin, pp 149–164

Lucas K (2012) Transport and social exclusion: where are we now? Transp Policy 20:105–113. https://doi.org/10.1016/j.tranpol.2012.01.013

Lucas K, Van Wee B, Maat K (2016) A method to evaluate equitable accessibility: combining ethical theories and accessibility-based approaches. Transportation 43(3):473–490. https://doi.org/10.1007/s11116-015-9585-2

Martens K (2012) Justice in transport as justice in accessibility: applying Walzer's 'Spheres of Justice' to the transport sector. Transportation 39(6):1035–1053. https://doi.org/10.1007/s11116-012-9388-7

Martens K (2017a) Transport justice: designing fair transportation systems. Routledge, New York, London

Martens K (2017b) Why accessibility measurement is not merely an option, but an absolute necessity. In: Punto N, Hull A (eds) Accessibility tools and their applications. Routledge, New York, London

Martens K, Golub A (2012) A justice- theoretic exploration of accessibility measures. In: Geurs KT, Krizek KJ, Reggiani A (eds) Accessibility analysis and transport planning: challenges for Europe and North America. Elgar, Celtenham, pp 195–210

Mokhtarian PL, Cao C (2004) TTB or not TTB, that is the question: a review and analysis of the empirical literature on travel time (and money) budgets. Transp Res Part A Policy Pract 38(9):643–675. https://doi.org/10.1016/j.tra.2003.12.004

Nielsen (2004) Universo de Establecimientos Detallistas. Bogotá

Preston J, Rajé F (2007) Accessibility, mobility and transport-related social exclusion. J Transp Geogr 15(3):151–160. https://doi.org/10.1016/j.jtrangeo.2006.05.002

Pucci P, Vecchio G, Bocchimuzzi L, Lanza G (2019) Inequalities in job-related accessibility: testing an evaluative approach and its policy relevance in Buenos Aires. Appl Geogr 107:1–11. https://doi.org/10.1016/j.apgeog.2019.04.002

Rodríguez DA, Targa F (2004) Value of accessibility to Bogotá' s bus rapid transit system. Transp Rev 24(5):587–610. https://doi.org/10.1080/0144164042000195081

Stanley J, Vella-Brodrick D (2009) The usefulness of social exclusion to inform social policy in transport. Transp Policy 16(3):90–96. https://doi.org/10.1016/j.tranpol.2009.02.003

Urry J (2007) Mobilities. Polity Press, Cambridge

Van Wee B (2011) Transport and ethics: ethics and the evaluation of transport policies and projects. Elgar, Celtenham

Van Wee B (2013) The traffic and transport system and effects on accessibility, the environment and safety. In: Van Wee B, Annema JA, Banister D (eds) The transport system and transport policy. Elgar, Celtenham, pp 4–18

Van Wee B (2016) Accessible accessibility research challenges. J Transp Geogr 51:9–16. https://doi.org/10.1016/j.jtrangeo.2015.10.018

Van Wee B, Geurs K (2011) Discussing equity and social exclusion in accessibility evaluations. Eur J Transp Infrastruct Res 11(4):350–367

Vecchio G (2017) Democracy on the move? Bogotá's urban transport strategies and the access to the city. City Territ Archit 4(15):1–15. https://doi.org/10.1186/s40410-017-0071-3

Vecchio G (2018a) Movilidades periféricas en Bogotá: hacia un nuevo paradigma. Quid 16(10):182–209

Vecchio G (2018b) Producing opportunities together: sharing-based policy approaches for marginal mobilities in Bogotá. Urban Sci 2(3):54. https://doi.org/10.3390/urbansci2030054

World Bank (2002) Cities on the move: world bank urban transport strategy review. The World Bank, Washington. https://doi.org/10.1596/0-8213-5148-6

Zahavi Y (1974) Traveltime budgets and mobility in urban areas. Final Report, Washington

Chapter 3
Emerging Mobilities: New Practices, New Needs

Abstract This chapter introduces new research findings on mobility practices in Italy that suggest a transformative nexus for explaining the role of mobility in contemporary cities. Superseding simplified interpretations of mobility as movement through space, this chapter describes emerging mobility practices and their temporal space variability to highlight socio-economic and lifestyle transformations, with particular attention on long-distance daily commuting (LDC). As an emerging form of work-related mobility, LDC is the result of the combined effects of an evolving labour market, that requires more and more flexibility while also subjected to increasing degrees of uncertainty, as well as of the territory and the transport and communication networks that allow the lengthening of travel. The interest in these mobility practices concerns the conditions that determine them, the consequences in lifestyle and in the uses of a territory and its networks. Through a sequence of quantitative analyses, supported by complementary and qualitative surveys and several interviews, the chapter analyses the emerging needs, times and conditions of using spaces and networks. It also examines the intensity of interactions activated by these practices that question traditional mobility services and provisions, and generate emerging new goods and services.

Keywords Long distance commuters · Mobility practices · Transport services · Reversibility

3.1 Introduction: Why We Deal with Emerging Mobilities

The socio-economic and territorial transformations in contemporary cities have increased fragmented and diversified mobility for individuals. They have introduced a significant complexity in 'trip chaining'—that is, the way in which people use only one trip to reach different places for different purposes—which is matched by a decrease in the weight of 'systematic trips' (home-work/study) in relation to the entire amount of travel.

This chapter was authored by Paola Pucci.

Long-distance commuting, job-related mobilities at a regional, national and international scale, multi-locality and mobility in atypical time slots are some of these emerging work-related forms that make daily mobility ever more complex. These emerging work-related mobilities—due to their increasingly spatial and temporal variability—are no longer comparable to traditional categories of commuting. They use space and services in new ways, according to specific times, preferences and constraints that are linked to long daily travel times. In these practices, we can read the way in which changes in working patterns impact mobility habits in terms of time and frequency devoted to displacements, distances covered and activities carried out during travel.

The interest in focusing on these emerging forms of mobility, still quantitatively insignificant compared to daily commuting flows, is twofold. On the one hand, by analysing them we can highlight the impact of socio-economic and life-styles transformations, as they result from the combined effects of labour market evolution, as well as of the ownership of the territory, and transport and communication networks (Kaufmann 2005). In this way, they work as a knowledge tool (an 'analyseur' in the words of Bourdin 2005, p. 17), useful for describing the socio-spatial-temporal transformations in the labour market and in work-programs in contemporary society. On the other hand, based on their space-time complexity, they contribute to questioning two key principles of the traditional 'utilitarian approach' in transport planning: the concept of travel as a derived demand and the understanding of users' objective as travel cost minimization. This is because, in these mobilities, the increase in travel time budgets is no longer experienced as wasted time, but increasingly as a social and productive time, used to work, read books and newspapers or chat and meet with colleagues and friends on the train or at the station. The same travel choices are guided by much more complex criteria than time cost minimization: they involve a plurality of aspects linked to both individual (preferences and norms, social ties, household composition) and external factors (transport supply, housing market, labour market).

In these mobilities, contrary to the idea of disutility in transport research, which assumes travel time is a cost to be minimised, travel time has been shown to be valued for a variety of reasons (Lyons and Urry 2005; Jain and Lyons 2008). From this basis, studying emerging mobilities allows us to describe more than displacements and their intensity just in terms of flows, but more importantly detect the combined effects of socio-economical and lifestyles changes, individuals' aims as well as territorial and infrastructure transformations (Pucci 2016).

The chapter focuses on a specific form of emerging mobility, that of long distance commuters (LDCs). The aim is to investigate the relevance of these practices, the reasons behind the origin of these mobilities and their impacts on daily life. After introducing the primary existing research on long distance commuters and its interpretations of these mobilities, the chapter then analyses the case of long distance commuters in the Milan Urban Region (Northern Italy). It does so through an analytical approach that matches qualitative and quantitative data sets that describe these mobilities in their spatial and behavioural dimensions. By analysing the emerging needs, times and conditions of using spaces and networks, as well as the intensity of interactions activated by these practices that question traditional mobility services and provisions, the conclusions identify three main profiles of LDCs and possible relevant policy measures for this mobile part of the population. The findings contribute to informing services offered in stations and inside vehicles, the equipment in new spaces of mobility as well as interventions beyond infrastructural provisions.

3.2 Long Distance Commuters as Emerging Mobility. Research Evidence

The relevance of emerging mobilities in describing new life-styles and work constraints has been described by several authors who pay attention to different aspects of these practices (Schneider and Meil 2008; Ralph 2014; Viry and Kaufmann 2015; Bissell et al. 2017). They refer to multi-locality experiences such as overnighters (those spending at least 60 overnight stays away from their home during a 12 months period for occupational reasons), recent re-locators, as well as people in long distance relationships and long distance daily commuters. Among these emerging mobility practices, long distance commuters (LDCs) appear as a phenomenon of real interest in the European context. During his/her working life, one in two people, if not personally then at least indirectly as a family member, has experienced a period of long distance commuting, though usually for a limited period in his/her professional life (Viry and Kaufmann 2015).

Even if LDCs have been the focus of only a small body of current research, we find a set of shared reasons that originate this mobility, namely fast travel modes, efficient intermodal connections and the spreading of urban populations in peri-urban areas (Ralph 2014; Viry and Kaufmann 2015; Bissell et al. 2017). At the same time, these studies do not offer converging interpretations on the impacts of these mobilities on daily-life, the motivations of people involved in this daily-travel nor on the attributes and criteria necessary to define long distances commuters.

For Schneider and Meil (2008), LDCs are part of other emerging work-related long-distance mobilities and they involve workers who spend more than two hours at least three times a week, to reach their workplace, in order to avoid residential mobility (p. 38). LDCs result from the combined effects of labour market evolution and a familiarity with the territory and its transport and communication networks that lengthen travel (Viry and Kaufmann 2015). Commuting for long distances allows these workers to have the sedentary lifestyle they seek, thanks to the performance of transport networks (speed of connection) that can "combine, in everyday life, places of activities that were before spatially impossible to assemble" (Vincent-Geslin and Ortar 2012, p. 40). Based on a quali-quantitative survey on a significant sample of highly mobile people in Europe,[1] Viry and Kaufmann (2015) observed that LDCs

---

[1] The EU research "Job mobilities and Family Lives in Europe" (http://www.jobmob-andfamlives. eu/) investigated the job related high mobilities of a sample of 7220 people in 2007 in six European countries (Germany, France, Spain, Switzerland, Poland and Belgium). This survey has been updated in 2011 only in Germany, France, Spain, Switzerland.

experience their extensive travel as reversible, more typically as a spatial reversibility, rather than a social one. This is because, according to Viry and Kaufmann (2015, p. 10), two dimensions define reversible mobility: "spatial and temporal reversibility, using travel time to stay in touch with friends and family, using travel speed to be physically present with friends and relatives as much as possible; (and) existential and relational reversibility, compensating for absence by maintaining distant relationship, limiting the impact of absence and limiting the contact with unfamiliar places and unknown people by developing routines".

The concept of reversibility,[2] according to these authors, becomes a key element in explaining the life-style conditions of LDCs. For them, the compression of time and space provides more leeway for individuals and is accompanied by a growing mobility injunction that tends to compel the social movement of individuals (Vincent-Geslin et al. 2016, p. 47).

This condition has also been highlighted by Schneider and Meil (2008), according to whom LDC is the most widespread form of recurring high mobility, involving from 11 to 15% of the workforce in Switzerland, France and Germany (p. 33).

For Bissell et al. (2017), long-distance commuting practices are "routine forms of mobility that go beyond the local and urban level, but that are not quite as broad-ranging as international travel" (p. 795). The authors studied this phenomenon through an ethnographical approach focused on the experiences of 'supercommuters', especially as they pertain to the 'intensity' of their lives and mobility experiences. The concept of intensity, according to the authors, can help to address "the singular and situated dimensions of mobility experiences that often get sidelined in more generalising socio-psychological explanations for commuting practices" (p. 796). In doing so, and through exploring three fieldworks in Australia, Canada and Denmark,[3] Bissell et al. (2017) found that participants in their research experienced an 'embodied knowledge' that emerged through a "complex interplay with knowledge, technologies, embodied sensations, and the travelled environment". In their interpretation, "the supercommuter accumulates a number of bodily capacities to affect and be affected by these systems and infrastructures" (p. 809).

Reconstructing the societal shifts that may have propelled long-distance commuting, Ralph (2014) examined a particular subset of the so-called Eurostar population: the 'Euro-commuters'.[4] According to the author, these are "EU citizens moving

---

[2]According to Kaufmann (2002, p. 25) "reversibility and irreversibility must be considered as ideal-types in that, forms of mobility are never purely reversible or irreversible".

[3]In Australia the sample is composed by 53 commuters in Sydney, routinely travelling 35 min each way, who responded to a newspaper advert calling for participants who experience stress as part of their everyday mobile life. In Canada, the sample is made by 18 floatplane pilots working as a unique commuter for commercial companies based in coastal British Columbia. In Denmark the sample are jet supercommuters defined as frequent fliers of a 45 min daily Scandinavian Airlines flight connecting the cities of Aalborg and Copenhagen (Bissell et al. 2017).

[4]As mentioned by Ralph (2014), the term is a play on Adrian Favell's definition of 'Eurostars', describing the EU citizens who live outside their home country in another member state. According to Favell (2008) these people are a pioneering group of EU citizens in an ever-more integrating Europe.

across EU borders routinely" and chronically commuting back and forth. If the reasons for an increase in these European commuting practices can be explained through the removal of formal labour market and mobility barriers for EU citizens, as well as low cost air fares and high speed transport lines, "Euro-commuters differ from Eurostars in that their migration is, from the outset, planned as a more transient project, rather than as a permanent settlement in the host destination" (Ralph 2014, p. 3). Focusing on the daily lives of a sample of these Euro-commuters who live between their home and another host European country, Ralph (2014) analysed how these people conduct their family life, friendships and careers to distinguish, among them, three typologies strongly inflected by gender: "survivors who undertake this mobility primarily for livelihood purposes; thrivers for the sake of optimizing lifestyles; strivers do so in the service of upward career mobility" (p. 34). These three profiles, depending on age, family formation stage, education and specialized training, influence individual's mobility capital and are affected by differing primary motivations in undertaking the shuttle-like migration between their country and other EU destinations. Their reasons can range from contracting economy, falling standards of living and spiralling mortgage-related debts at home to a dysfunctional housing market, as well as improving economic status, career promotional considerations and advancing their professional occupational status. Becoming an alternative to more conventional forms of migration, according to Ralph (2014), this kind of transnational circular mobility is nowadays a prospect for ordinary European professionals to consider as part of their attempts to achieve some sort of durable work-life balance.

   In one of the first studies carried out on the phenomenon, Öhman and Lindgren (2003) explored the determinants of long-distance commuting as compared to shorter commuting distances, considering individuals with a Euclidean distance of 200 km or more between their housing and work place. The research studied the Swedish population and long-distance commuting emerged as a mobility strategy to manage the growing distances between home and work places. It revealed that mobility strategies and the lifestyle of long-distance commuters suggests a diversity of driving forces affecting individual and household preferences along with structural factors. The most influent factors have been summarized into seven categories representing personal characteristics, preferences and norms, household composition, social ties, labour market conditions, housing market conditions and transportation (Öhman and Lindgren 2003, p. 11).

   If the attributes and criteria in defining LDC are not unequivocal in the above mentioned studies, in which LDC can be detected by considering both travel times and spatial distances covered to reach work places, however, "the crossing of space, in terms of speed and distance, should not be used as an indicator to describe (this) mobility" (Kaufmann 2016, pp. 40–41). These mobilities in fact describe more than displacements: they derive from the lifestyles and plans of actors and also depend on obligations, constraints or windows of opportunity open to us. Through these emerging practices, it is possible to analyse the ways in which resources (tangible and intangible, physical and personal) and individual goals are combined and, on the other hand, recognize different ways of using travel time as a 'productive time' for working, connecting socially or relaxing, "becoming a social time in its own right" (Kaufmann 2016, pp. 40–41).

Research has aimed to understand the complexity of these mobility practices while dealing with the structural and influent factors at the origin of these displacements. To cope with the consequences of these practices on the uses of the transport networks and their services, all the examined studies have used qualitative approaches and surveyed a sample of the population experiencing long distance commuting. Since these emerging mobilities need to be analysed in their spatio-temporal complexity, quantitative data can detect the intensity and spatial dimensions of this phenomenon. Together with the networks that support these mobilities, quantitative studies need to be combined with qualitative approaches that analyse resources, preferences, constrains, social ties and individual goals, intercepting also the needs expressed and goods generated by the same mobility practices. This has been the approach undertaken to investigate LDC in the Milan Urban Region.

3.3   Long Distance Commuters in Italy. A Focus on the Milan Urban Region

Analysing work-related displacements in Italy with quantitative data sources,[5] new strategies for reconciling private and professional life emerge. These can be seen through new forms of daily mobility that include long-distance commuting; job-related mobilities at a regional, national and international scale as well as multi-locality and mobility in atypical time slots (Pucci 2017, 2018). Among them, long distance commuting[6] for work-reasons in Italy has increased significantly, especially in women's mobility, representing 36.7% of the long distances displacements exceeding 150 km. For work-related displacements exceeding 150 km per day, some evidence shows a strong relationship between the increasing trend of this travel and the existence of infrastructure corridors with high-speed lines that provide a good transport supply for long commuting.

Considering the transport means, the relevance of trains increases for the displacements between Turin and Milan, Milan and Rome and Rome and Naples served by the high speed lines, even if the car is the most used means of transport in journeys exceeding 150 km (65% of total LDC displacements). This mode is preferred above all others, when the radius of displacement is within 300 km, while for travel up to 300 km, in addition to the train (22%), plane travel also emerges (15%) as an option. Taking into account work-related displacements up to 75 km one way, two

---

[5]The source is the national census by Istat, available at the municipality level for all the Italian territory. Istat's census (2001–2011) provides data only on Origin and Destination of commuters' flow for study and work reasons, with additional information on modal share and time of displacement.

[6]In our research, we selected LDCs by considering travel distances over 75 km one way. The distances have been calculated using the graph of the infrastructure network. They are therefore geographical distances between an origin (home) and a destination (workplace).

Fig. 3.1 Long distance
commuters in Italy:
displacements up to 150 km

different situations emerge at the national level: LDC trips are strongly polarized
in the main cities in Southern Italy with particular evidence in Rome (with strong
origins in Naples), as well as in Bari and Palermo. Meanwhile, LDCs destinations
are more equitably distributed in Northern Italy, where a polycentric network centred
on Milan emerges and is characterised by the attractiveness of other polarities, like
Vicenza and Padova. This dynamic is confirmed also for trips exceeding 150 km,
even if the attractiveness of Turin is weaker. In the displacements up to 300 km,
Milan and Rome play a relevant role as the main attractive poles (Fig. 3.1).

These are the most relevant trends detected from the quantitative data sources
available at the national level. The elaboration of quantitative data highlights this
trend as an emerging and growing phenomenon in Italy over the last 10 years. Though
it describes the spatial dimension and the networks that support this mobility, the data
does not offer specific indications for the reasons/habits/constrains of this mobility
practice. These dimensions play a relevant role in explaining the features of this
emerging mobility and understanding the reasons, needs and constraints relating to
long distances commuting. When discussing emerging mobilities, we refer to work-
related mobility practices where mobility, as a movement, is an 'epiphenomenon'
of wider transformations in the job market, thus it is capable of describing social,
spatial and temporal changes.

While quantitative and aggregated data contribute to bringing out the phenomenon
in its spatial distribution, they are not suitable for investigating the behavioural component (the experiences, capacities, needs) of this emerging mobility. Accordingly,

the approach used in this study integrates a sequence of quantitative analysis with complementary qualitative survey data, based on several interviews with a sample of LDCs. Through the quantitative data collected, the research analysed the movement component of the LDC phenomenon, in particular the spatial dimension and the networks that support this mobility. Meanwhile, the qualitative data has deepened the behavioural components. In particular, a qualitative approach has been finalised to analyse the ways in which resources (tangible and intangible, physical and personal) and individual goals are combined, intercepting needs expressed and goods generated by these mobility practices.

This in-depth study that combines quantitative with qualitative approaches focuses on the daily mobility practices in a dense metropolitan area: the Milan Urban Region (Northern Italy). The choice to investigate this territory is linked to its complex urban settlement, where the multi-directional daily mobility reflects a complex network of the relationships and places densely used that follow a non-hierarchical urban structure. In this territory, the geography of the movements defines an extensive area which goes beyond the institutional borders of the metropolitan city of Milan (*Città Metropolitana di Milano*), an element that challenges institutional governance systems, ineffective in governing the current mobility dynamics. As the focus on a dense metropolitan area where congestion distorts the measurement of travel time, travel distances over 75 km between an origin (home) and a destination (workplace) were assumed as a point of reference for selecting and investigating LDC.

The approach accompanies quantitative and qualitative surveys that contribute to the discovery of the time-space variabilities of these emerging mobilities. It also captures changes in work patterns and life-style that would not emerge statistically and tackles the practical issues they raise.

To detect the intensity and spatial dimensions of LDC, together with the networks that support this mobility, quantitative data[7] has been processed to investigate the size and trends of this form of mobility for the Milan Urban Region. The study also examines which territories have been most affected by this phenomenon along with their features in terms of transport supply and services, the socio-economical profiles of commuters, as well as how much time LDCs dedicate to travel each day.

Data shows an increase in the work-related displacements longer than 150 km by 43% from 2001 to 2011, yet they represent only 1% of total displacements in the Milan Urban Region (Fig. 3.2).

The average long distance commuter is male (80%), aged 30–59, employed and commutes each day from his home to workplace using a private car. Car usage prevails despite the fact that LDCs are concentrated in urban centres of medium-large size and these municipalities are served by a railway station and distributed along the lines of the regional railway system (Fig. 3.3).

---

[7]The sources are the national census by Istat and the Lombardy O/D matrix. Istat's census (2001–2011) provides data on commuters' flow for study and work reasons (O/D, modal share and time of displacement); O/D matrix by Regione Lombardia is a survey led in 2002 and 2014, on all daily displacements (reasons, modal share, time of displacement, professional profile, gender, age).

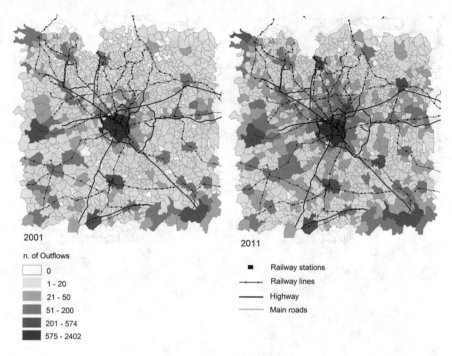

2001

2011

n. of Outflows

☐	0
	1 - 20
	21 - 50
	51 - 200
	201 - 574
■	575 - 2402

- ■ Railway stations
- ├─┼─ Railway lines
- ── Highway
- ── Main roads

Fig. 3.2 The geographies of LDCs

Results show that long-distance commuters travel primarily to reach their work-place, although in the last ten years this trend has begun to change in favour of business travel (meeting clients), in part due to the increase in freelance work following the 2008 economic crisis.

To understand the reasons for being a long distance commuter, the research inves-tigated the activities carried out during travel and the analysis highlights the emerging needs in terms of spaces, equipment and services useful for these practices. An ethno-graphic approach, based on semi-structured interviews, travel-along and mapping,[8] has been carried out on a selected sample of the LDCs.

The ethnographic approach has been finalised to investigate not only the reasons for commuting long distances, but also the "individual functional space" (Lévy 1994, p. 241) of these commuters more in depth, considering:

- the differentiated relationships with LDCs' home spaces. These refer to social ties including friends, family, colleagues, household composition, housing ownership, preferences and norms;
- the activities carried out during the period of travel and the uses of the spaces of movement. These involve reading books and newspapers or chatting with friends,

---

[8]The approach has been experienced in Vendemmia's PhD thesis (2015) and in Akhond's Master thesis (2017), both developed in the Politecnico di Milano. See also Vendemmia (2017).

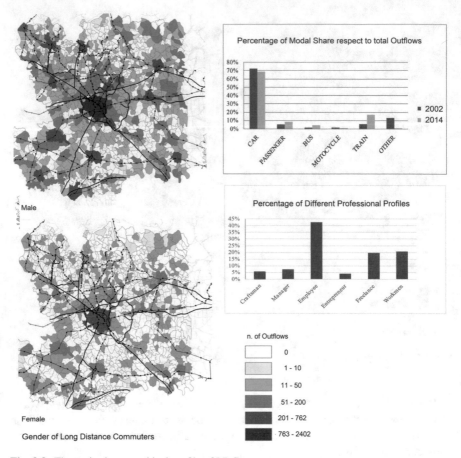

Fig. 3.3 The socio-demographical profile of LDCs

meeting colleagues and friends on the train or at the station and working on their laptop;

- the emerging needs raised by LDCs. These include minimizing time constraints, equipping transport spaces with comfortable chairs, electric plugs useful for short-term work, free Wi-Fi connection and meeting places and even equipping the train coach as a working space.

The sample of people with mobility practices definable as long distance or commuting for an extended period of time has been selected using the snowball technique. It was composed of 15 highly mobile persons, aged between 31 and 48, with differing social profiles. The sample is composed of 6 women and 9 men, 6 of whom have children; 6 of whom are married and 7 with a common law marriage. They are all highly educated, holding either a bachelor, master or PhD degree. 6 currently have an open-ended contract, 2 are free-lance professionals, 4 have a fixed term contract and 3 have a per-project contract. The analysis of the 15 individuals focused not only

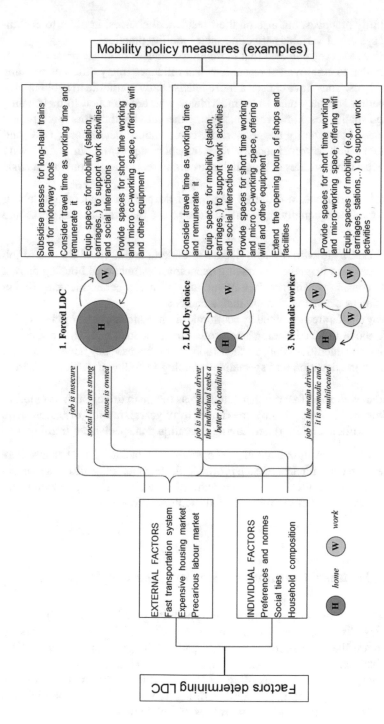

Fig. 3.4 Profiles of LDCs and possible relevant policy measures. *Source* Pucci and Vecchio (2019)

on their mobility practices, but also on their residential choices, in order to explain the complex system of variables affecting their mobility practices.

The qualitative survey detected that these long distances mobility practices originated mainly due to employment reasons, which influence many different domains of the LDCs' lives. They have chosen a more intense daily commute to avoid migration or relocation and they tend to be individuals who are attached to their habits and social and spatial ties. However, considering the individual factors (preferences and norms, social and family ties, household composition) determining LDC and combining them with 'external factors' (fast transport supply, labour and housing markets), at least three profiles of long distance commuters emerge from the analysis (Fig. 3.4).

These profiles translate to different degrees of adaptation to social and labour market constraints and equally different abilities to use the available networks and services:

- the forced LDCs are workers who decide to undertake this mobility due to job insecurity resulting from the contracting economy, dysfunctional housing market and family constrains. Their main motivation is primarily economic; they live in this condition as a constraint, more than a choice;
- the LDCs by choice are not forced by economic circumstances, but they choose this way of life above all for advancing their professional occupational status and career promotional considerations enforce their mobility practices. They have often spent several years in education and specialized training to follow a particular 'career path';
- the nomadic workers with a multi-located job as the main driver of their mobility habits. They are also specified by age (under forty years), family formation stage (more often without children) and having education and specialized training.

These profiles can be significant also to operationally tackle LDCs because they highlight the main determinants of each profile. This understanding allows a rethinking of the services to be offered in stations and inside vehicles, the organization and equipment of new spaces of mobility, as well as interventions beyond infrastructure provisions.

3.4 Long Distance Commuters: Beyond Just Displacement

Studying LDC as an emerging work-related mobility practice allows for a reading beyond just displacements. It underlines the way in which changes in working patterns increase the complexity of mobility patterns in terms of frequency, time devoted to displacements, distances covered, as well as the implications in the daily-life of this type of mobility. According to Bourdin (2005), this mobility clearly represents a knowledge tool, useful for describing transformation in urban life and work-programs in identifying different profiles of mobile people.

To follow the relevance of this mobility in its space-time variability, as well as the reasons, behaviours and constrains related to this mobility practice, the need emerges for analytical tool-boxes capable of combining 'small data', including qualitative data, together with big data, in order to overcome the limitations of the latter (Kitchin and Lauriault 2015). It is not a matter of distinguishing between qualitative and quantitative approaches, but of addressing the analytical approach in a useful way to explore the dimensions characterizing this mobility as a movement in space, in which the behavioural components are both relevant to address interpretations of this phenomenon and are useful for improving the effectiveness of mobility policies. This requires an analysis of the intensity and spatial dimensions of this phenomenon, the networks that support these mobilities, together with the resources (tangible and intangible, physical and personal), constraints and individual goals that originate this mobility practice.

An important contribution in this sense comes from digital data sources, generated by users and which can contribute to a better understanding of mobility behaviours. The geographical scale and temporal dimensions are almost completely absent from conventional surveys, despite being one of the most significant elements that characterize not only this type of mobility, but contemporary cities and urban regions, in general. Potentially, digital data could lead to defining more user centred policies. After a decade of experimentations (Ratti et al. 2006; Kwan et al. 2007; Reades et al. 2007; Järv et al. 2014; Blondel et al. 2015; Pucci et al. 2015), it is indeed possible to state that mobile phone data with digital data and ICT have improved understandings of urban practices and mobility in many respects. This data can simultaneously overcome the limitations of the detection latency time typical of traditional data sources and take advantage of the ubiquity of mobile phone networks and the pervasive diffusion of mobile devices. In addition, they provide a 'longitudinal perspective' on the variability in human travel activities (Järv et al. 2014), thus validly complementing traditional research methods.

These methods improve the understanding of these practices, as well as help in detecting emerging needs for addressing more effective policy, better equipped to assess demands and initiatives in the transport market: rather than intervening on transport infrastructures, LDCs could benefit most from improvements to services and facilities that would support their work activities and social interactions during the travel (Fig. 3.4).

In the interviews conducted, the 'productivity' of travel clearly emerges because the increase in travel time budgets is no longer experienced as wasted time, but it has increasingly become a social time in its own right. LDCs work, read, use their laptops and digital devices to interact via internet, meet colleagues and friends during their travel, and due to this possibility, the choice of means of transport is not necessarily dictated by economic convenience, but by the quality of the activities that the means of transport allows users to carry out during the journey.

LDCs in the Milan Urban Region reveal the effect of the evolution of the labour market on these mobility practices, as well as the features of the territory, transport and communication networks, allowing a minimization of physical distances. In doing so, overcoming an initial and stereotypical idea of LDCs as élites with mobile lives, the empirical evidence shows that often long daily travel is a necessary and unavoidable condition to meet economic and family needs and/or housing inequalities. For these reasons, this mobility is an expression of a 'rooting in the fluidity'.

References

Akhond R (2017) Long-distance commuting, new emerging mobility practices in Milan urban region. Master thesis, Politecnico di Milano

Bissell D, Vannini P, Jensen OB (2017) Intensities of mobility: kinetic energy, commotion and qualities of supercommuting. Mobilities 12(6):795–812. https://doi.org/10.1080/17450101.2016.1243935

Blondel V, Decuyper A, Krings G (2015) A survey of results on mobile phone datasets analysis. Arxiv. http://arxiv.org/abs/1502.03406

Bourdin A (2005) Les mobilités et le programme de la sociologie. Cah Int Sociol 118(1):5–21. https://doi.org/10.3917/cis.118.0005

Favell A (2008) Eurostars and Eurocities. Blackwell, Oxford

Jain J, Lyons G (2008) The gift of travel time. J Transp Geogr 16:81–89

Järv O, Ahas R, Witlox F (2014) Understanding monthly variability in human activity spaces: a twelve-month study using mobile phone call detail records. Transp Res Part C: Emer 38(1):122–135

Kaufmann V (2002) Re-thinking mobility. Ashgate, Farnham

Kaufmann V (2005) Mobilités et réversibilités: vers des sociétés plus fluides? Cah Int Sociol 118(1):119–135. https://doi.org/10.3917/cis.118.0119

Kaufmann V (2016) Putting territory to the test of reversibility. In: Pucci P, Colleoni M (eds) Understanding mobilities for designing contemporary cities. Springer, Heidelberg, New York, Dordrecht, London, pp 35–48

Kitchin R, Lauriault TP (2015) Small data in the era of big data. Geo J 80(4):463–475. https://doi.org/10.1007/s10708-014-9601-7

Kwan MP, Dijst M, Schwanen T (2007) The interaction between ICT and human activity-travel behaviour. Transp Res Part A Policy Pract 41(2):121–124. https://doi.org/10.1016/j.tra.2006.02.002

Lévy J (1994) L'Espace légitime. Presses de Sciences Po, Paris

Lyons G, Urry J (2005) Travel time use in the information age. Transp Res A 39:257–276

Öhman M, Lindgren U (2003) Who are the long-distance commuters? Patterns and driving forces in Sweden. Cybergeo Eur J Geogr 243:1–23. https://doi.org/10.4000/cybergeo.4118

Pucci P (2016) Mobility practices as a knowledge and design tool for urban policy. In: Pucci P, Colleoni M (eds) Understanding mobilities for designing contemporary cities. Springer, Berlin, pp 3–22

Pucci P (2017) Metabolismo e mobilità: se i flussi si fanno tracce. In: Balducci A, Fedeli V, Curci F (eds) Metabolismo e regionalizzazione dell'urbano. Esplorazioni nella regione urbana milanese. Guerini e Associati, Milano, pp 115–127

Pucci P (2018) Post-metropoli: una città in movimento. In: Balducci A, Fedeli V, Curci F (eds) Italia post-metropolitana: scenari di innovazione per una nuova questione urbana. Guerini e Associati, Milano, pp 171–196

Pucci P, Vecchio G (2019) Trespassing for mobilities. Operational directions for addressing mobile lives. J Transp Geogr (Manuscript submitted for publication)

Pucci P, Manfredini F, Tagliolato P (2015) Mapping urban practices through mobile phone data. Springer, Heidelberg, New York, Dordrecht, London

Ralph D (2014) Work, family and commuting in Europe: the lives of Euro-commuters. Palgrave, London

Ratti C, Frenchman D, Pulselli RM, Williams S (2006) Mobile landscapes: using location data from cell phones for urban analysis. Environ Plann B Plann Des 33:727–748. https://doi.org/10.1068/b32047

Reades J, Calabrese F, Sevtsuk A, Ratti C (2007) Cellular census: explorations in urban data collection. IEEE Pervasive Comput 6(3):30–38. https://doi.org/10.1109/MPRV.2007.53

Schneider NF, Meil G (eds) (2008) Mobile living across Europe I: relevance and diversity of job-related spatial mobility in six European countries. Barbara Budrich Publishers, Opladen

Vendemmia B (2015) What spaces for highly mobile people? Analyzing emerging practices of mobility in Italy. PhD thesis, Politecnico di Milano

Vendemmia B (2017) Are emerging mobility practices changing our cities? A close look at the Italian case. In: Hartmann-Petersen K, Freudendal-Pedersen M, Perez Fjalland EL (eds) Experiencing networked urban mobilities. Routledge, London

Vincent-Geslin S, Ortar N (2012) De la mobilitè au racines. In: Vincent-Geslin S, Kaufmann V (eds) Mobilité sans racine. Plus loin, plus vite… plus mobiles? Descartes and Cie, Paris

Vincent-Geslin S, Ravalet E, Kaufmann V (2016) Des liens aux lieux: l'appropriation des lieux dans les Grandes mobilités de travail. Espaces et Société 164–165:179–194

Viry G, Kaufmann V (2015) High mobility in Europe. Work and personal life. Palgrave Macmillan, London

Chapter 4
Big Data: Hidden Challenges for a Fair Mobility Planning

Data are both social and material, and they do not merely
represent the world but actively produce it.

(Kitchin 2014b, p. 226)

Abstract Big data provides unprecedented opportunities for understanding and planning urban mobility. Big data makes huge amounts of information available with a high level of detail, potentially precisely depicting overall macrotrends while also detecting micropractices that elude traditional investigative approaches. Despite this unprecedented insight on mobility issues, big data is not the ultimate solution for dealing with urban mobility, especially when considering social dimensions. The chapter intends to discuss at what conditions big data may contribute to mobility planning and policy approaches that may effectively enable individuals and their opportunities. After introducing the concept of big data and its increasing relevance, the significance for urban policy of such a knowledge source is briefly presented. The discussion then moves to four critical issues that question the contribution of big data to enabling mobilities. These include the representativeness of the information provided by big data; the interpretative issues associated with understanding mobilities through manifold digital technologies; the differentiated individual ability to produce information through portable devices and emerging actors who operate through big data and influence urban mobility dynamics in unforeseen ways. These dimensions highlight three forms of partiality that affect the completeness, the neutrality and the usability of big data in relation to mobility issues, leading to a call for a critical usage of big data. This approach can contribute to enabling mobilities thanks to the enriched information it provides, without delegating to it the responsibility of defining urban problems and solutions.

Keywords Big data · Mobility · Transport planning · Urban policy

This chapter was authored by Giovanni Vecchio.

© The Author(s), under exclusive license to Springer Nature Switzerland AG 2019   43
P. Pucci and G. Vecchio, *Enabling Mobilities*, PoliMI SpringerBriefs,
https://doi.org/10.1007/978-3-030-19581-6_4

4.1 Introduction

Big data is an increasingly pervasive presence in contemporary cities and societies. On the one hand, the number of devices able to track movements, preferences and consumption is growing, as well as their diffusion among users. On the other hand, the analyses that take advantage of such data are becoming each day more sophisticated and represent urban phenomena with high levels of precision. As for many features of the wide wave of technological innovations affecting urban settings, big data is controversial. There is huge enthusiasm for the analytical and operational opportunities it provides, as well as concern for the forms of control and oppression it may support (Greenfield 2017, Chap. 8). The pervasiveness of technology results in ambiguous potential effects on cities and their inhabitants, making it thus necessary to considerately engage with big data and the outcomes resulting from its growing relevance.

Big data also provides new opportunities for understanding and planning urban mobility, and it requires a change in the mainstream approaches to transport issues (Milne and Watling 2019). Big data, in fact, makes huge amounts of information available with a high level of detail, potentially providing both precise overall depictions of macrotrends while simultaneously detecting micropractices that elude traditional investigative approaches. The mobile technologies that support our increasingly mobile and connected lives (Elliott and Urry 2010) track our habits and collect the information that, aggregated, constitutes big data. At the same time, many innovative mobility services require information as a condition to access them and continuously collect information while operating. For example, it is the case of many sharing mobility initiatives to profile their users according to the zones and the hours in which they move. Big data thus emerges as a powerful but also risky tool for understanding and addressing mobility issues. The knowledge it provides and the operational opportunities it opens up are neither neutral nor exhaustive when considering mobility issues. Therefore, despite its unprecedented analytical insights, big data is not the ultimate solution for dealing with urban mobility, especially when considering its social dimensions.

The chapter intends to discuss under which conditions big data may contribute to mobility planning and policy approaches that may effectively enable individuals and their opportunities. After defining big data and how it is becoming increasingly relevant, the chapter briefly presents the urban policy significance of such sources of knowledge. Through a critical review of existing literature, the discussion then moves to four critical issues that question the contribution of big data to enabling mobilities. These include: the representativeness of the information provided by big data; the interpretative issues associated with understanding mobilities through manifold digital technologies; the differentiated individual ability to produce information through portable devices and emerging actors who operate through big data and influence urban mobility dynamics in unforeseen ways. These dimensions highlight three forms of partiality that affect the completeness, the neutrality and the usability of big data in relation to mobility issues. This leads to a call for a more critical usage of big data, one which can contribute to enabling mobilities thanks to the enriched information it provides, without delegating to it the responsibility of defining urban problems and solutions.

4.2 What Is Big Data?[1]

Despite its growing popularity, the term 'big data' does not have a single shared definition (Floridi 2012; Kitchin 2014c; Lovelace et al. 2016). Generally, big data is described by referring to the 'three V's': volume, velocity, and variety. Big data is "high volume, velocity and variety of data that demand cost-effective, innovative forms of processing for enhanced insight and decision making" (Gartner 2013). Three more V's could be added, considering that big data should also be veracious, variable and valuable (Gandomi and Haider 2015). Some key features of big data that also emerge from a review of the literature (Boyd and Crawford 2012; Kitchin and Dodge 2011; Marz and Warren 2015; Mayer-Schönberger and Cukier 2013; Zikopoulos and Eaton 2011) can be summarized as:

- *huge in volume*, consisting of terabytes or petabytes of data;
- *high in velocity*, being created in or near real-time;
- *diverse in variety*, being structured and unstructured in nature and often temporally and spatially referenced;
- *exhaustive in scope*, striving to capture entire populations or systems (n = all) or at least much larger sample sizes than would be employed in traditional, small data studies;
- *fine-grain in resolution*, aiming to be as detailed as possible and uniquely indexical in identification while still respecting individuals' privacy thanks to aggregation and anonymisation;
- *relational in nature*, containing common fields that enable the conjoining of different data sets;
- *flexible*, holding the traits of extensionality (new fields can be easily added) and scalability (can be rapidly expanded in size).

Big data includes different typologies of data, distinguishable according to the sources that provide them. Following Kitchin (2014c, p. 4), three categories of big data emerge:

- *directed data* or tracking data monitors the movement of individuals or physical objects subject to movement by humans, is generated by traditional 'forms of surveillance' that focus on people or places and implies a human operator;
- *automated data* is produced by automatic digital devices such as mobile phone data, capture systems and clickstream data that records how people navigate

---

[1]This section draws on Pucci et al. (forthcoming).

through a website or app; remotely sensed data generated by a variety of sensors and actuators as well as image data, particularly aerial and satellite images but also including land-based video images;

- *volunteered data*, deriving from search and social networking activities where citizens become 'sensors' and can contribute to the collection of geographic data, thus user generated. This includes: interactions with social media such as the posting of comments or the uploading of photos to social networking sites such as Facebook or Twitter as well as crowdsourcing data where users generate data and contribute to a common system, such as the generation of GPS-traces uploaded into OpenStreetMap to create a common, open mapping system (Goodchild 2007; Kitchin and Dodge 2011).

Big data needs to be examined in strict relationship to the algorithms and systems that allow it to be examined. Algorithms are fundamental in selecting relevant information out of the shapeless mass of big data and which then make sense of it. They are, for this reason, crucial to "provide a means to know what there is to know and how to know it, to participate in social and political discourse, and to familiarise ourselves with the publics in which we participate" (Gillespie et al. 2014). In this respect, the human contribution is crucial in shaping algorithms and consequently determining what information we consider and for what purposes. Algorithms may be a tool for justice but more often, intentionally or not, determine dynamics of control or exclusion (Cohen 2018). Therefore, calls for data-driven forms of policy in which 'data speaks for itself' or for an "automatic smart city understanding" (Villanueva et al. 2016, p. 1680) should consider the non-neutrality of algorithms. According to how these tools are shaped, very different understandings of spatial phenomena can be derived and, consequently, the operational measures deployed to deal with them.

Apart from the specific purposes to which big data may contribute, there are two main reflections regarding their usage in relation to urban issues. On the one hand, such data can contribute to examining urban phenomena in real time; on the other hand, it can enhance new forms of urban governance. Urban mobility practices offer an example in this sense. Big data, gathered from different sources (cameras, portable devices, electronic fare systems, etc.) can provide detailed and updated representations of how people move, considering both macrotrends and micropractices. In doing so, data updates and enriches traditional forms of knowledge of mobility; it can contribute to updating fundamental tools of transport planning, like origin/destination matrices and transport models (Noulas et al. 2012). Additionally, it can also analyse how the distribution of the population in cities changes through space and time (Ratti et al. 2006; Sevtsuk and Ratti 2010) and detect specific mobility behaviours and users' profiles (Bayir et al. 2010; Reades et al. 2007; Soto and Frías-Martínez 2011). The updated and precise knowledge of mobility fostered by big data allows, at least potentially, the deployment of faster and more effective responses to emerging urban issues. In addition to these aspects, big data can also enhance the ordinary management of services and infrastructures, contributing to defining and adapting real time interventions.

Big data is thus a primary component of a wider technology-based innovation in the ways we understand and address urban problems. While the growing diffusion of technologies is changing the lifestyles of people and their spatial reflections (for example, in the growingly complex forms in which they move), the data produced and gathered from using such technologies provides technicians and decision makers with an unprecedented amount of detailed information. This information is relevant not only for its quantity but also for its quality. It provides a richer understanding of how people move and how their mobility relates to other elements—for example, the activities that people undertake or the individual features that determine their needs, wants, preferences and perceptions. In doing so, big data can contribute to the ongoing shift in mobilities studies and transport planning which increasingly foster a wider understanding of mobility from a phenomenon understood as the sum of flows between manifold origins and destinations to a rich set of practices determined by individual and collective behaviours and habits. However, the richness and pervasiveness of big data can be overwhelming too, as "the picture we are left with is that of an urban fabric furiously siphoning up information, every square meter of seemingly banal sidewalk yielding so much data about its uses and its users that nobody quite yet knows what to do with it" (Greenfield 2017, p. 85). Because of this abundance, the relevance of big data is better understood by considering how it can intervene on shaping the responses to urban problems.

4.3 Big Data and Urban Mobility Policy[2]

From a policy perspective, big data provides a huge amount of information, offering a question-oriented knowledge. Depending on the question or issue faced by the subject using them, big data provides evidence by combining different datasets from different sources gathered for different purposes. While big data can contribute to addressing a wide range of mobility-related urban issues, from real-time traffic rerouting to accident management, here the focus is on urban policy. By policy, we refer to a deliberate system of principles that guide decision-making and lead to rational outcomes, identifying a course of action adopted for the sake of expediency, ease, etc. to solve a collective problem, that is a problem requiring public intervention. Policy can be seen as a cycle composed by different steps (Marsden and Reardon 2017), to which big data contributes differently. Here we choose to use policy cycle as a reference concept despite being aware of its limitations. Policy is not sequential in nature, rather it needs to be designed and continuously revised to take into account external conditions and adapt to their eventual change. Their effects are often indirect, diffuse and take time to appear and, finally, policy-making depends on politics, people, socio-economic factors as well as other previous and ongoing policies.

Problem setting. This is the first stage of the policy cycle, in which a collective problem is identified, defined and legitimised as such. This stage contributes to

---

[2]This section draws on Pucci et al. (forthcoming).

framing the issue to be faced, defining its main features and considering what ongoing policies (if any) are currently addressing it. In this phase, big data may contribute to understanding the current urban trends as well as to help assess existing policies. Real-time data that refers to the use of transport services and infrastructures is helpful in depicting current mobility flows by using a range of sources including traffic sensors, public transport passenger tracking and vehicle sharing services (Lu et al. 2015; Lv et al. 2014; Pelletier et al. 2011). Another crucial source is mobile phones, whose traffic data allows the mapping of the position of users and can contribute to profiling them. This data is relevant for mapping in real time the spatial and temporal variability of the distribution of populations in a territory, potentially contributing also to a dynamic redefinition of policy measures (Ahas et al. 2010; Pucci et al. 2015; Sevtsuk and Ratti 2010). Finally, social networking services provide significant information concerning the position of the users but also in conveying significant user-generated information related to ongoing activities, habits and sentiments (Chen and Schintler 2015; Liu et al. 2014; Pang and Lee 2008).

Policy formulation. In this stage of the policy cycle, alternative options of interventions are assessed in order to address the problem previously defined. In this phase, big data can contribute mainly to the definition and the evaluation of alternative scenarios, deriving from the adoption of one or more courses of action. The data contributing to the formulation of policy is similar to that used for problem setting, but its use is different. Data in fact contributes to defining alternative 'transport futures' with different degrees of possibility, plausibility and desirability (Banister and Hickman 2013). Scenario planning can use different methods to "create a set of the plausible futures", deriving from the adoption of alternative measures and the consequences these may generate (Amer et al. 2013, p. 25). Big data can help to define the outcomes of each scenario and therefore contribute to the choice of one specific course of action.

Policy design. During this stage of the policy cycle, the features of the policy are defined, depending on the pursued aims. In fact, the means to achieve different purposes can vary, leading therefore to manifold possible combinations of operational measures intended to achieve a given purpose. In this phase, the usage of the typologies of big data previously mentioned could be relevant. For example, these can promote forms of problem crowdsourcing (Brabham 2009), in which public participation is solicited for "harnessing collective intellect and creative solutions from networks of citizens in organized ways that serve the needs of planners" (Brabham 2009, p. 257). If the problem to be faced has a clear definition, big data can be useful for defining different courses of action having at disposal richer information.

Policy implementation. This stage of the cycle gives the policy its form and it becomes operational. Data can contribute to putting the policy into practice and reshaping it, if necessary, depending on the reactions of the setting or on the achieved results. In this stage, data contributes in at least five ways. First, it contributes to managing urban mobility issues in real time, allowing a "much more sophisticated, wider-scale, finer-grained, real-time understanding and control of urbanity" (Kitchin 2014c, p. 3). Second, subjects external to institutions—such as corporations and communities—may use data to develop their own initiatives addressing emerging urban mobility needs

(Vecchio and Tricarico 2019). Third, big data can help to improve the 'infostructure' of mobility by providing information that is updated, complete and personalized. This can contribute to enhanced individual mobility choices, better able to consider the wide range of available modal options (Schwanen 2015). Fourth, big data result from the wider involvement of citizens in the implementation of policy measures, for example through the provision of real-time feedback that can assess the obtained results and, in the case of negative reactions, contribute to the redesign of ongoing measures. Fifth, big data can be significant by simply being publically available to citizens. In this case, data is significant in providing information that allows people to observe how their city performs in relation to specific themes or issues, increasing the transparency and the accountability of the urban government. Moreover, making data open provides citizens with the opportunity of developing their own tools (such as websites and smartphone applications) to facilitate certain aspects of urban life, as demonstrated by the Singapore Live! project by the MIT SENSEable City Lab (Kloeckl et al. 2012).

Policy evaluation. The last stage of the policy cycle examines the results achieved with the implementation of a policy, estimating if the initial problem was solved, what expected or unexpected results were generated and what adjustments may be necessary for the policy to better perform in the future. Having defined relevant indicators to assess the results of a policy, big data can contribute to estimating the obtained outcomes and if the desired changes were obtained. Real time data can significantly contribute to observing, since its implementation, the first effects generated by the adopted policy while other data can provide relevant information—to be selected according to the faced issues and the outcomes to assess.

4.4   Can Big Data Enable Mobilities? Four Open Issues

Amongst the many issues raised by the pervasive diffusion of big data, the focus here is on four aspects that may question its contribution to enabling mobilities: these refer to the incompleteness of the information provided by big data, the interpretative issues regarding their usage, the differentiated abilities that individuals have for producing them and the new spaces of action they open for new actors in the field of mobility.

4.4.1   The Information of Big Data Is Not Complete

Big data is often praised for the amount of information it provides, but this information risks being incomplete. As the previously mentioned 3 Vs describe, the unprecedented advantage of big data is the result of a huge quantity of information, highly detailed, provided in real time and which is continuously updated. Also in relation to mobility issues, the mainstream approaches assume that big data is crucial to having a wider and deeper knowledge of urban issues, making it necessary to both expand the amount of available data and to refine the interpretative tools (such as algorithms) required to make use of it. Utopian scenarios emerge, such as that of a 'real time city' (Kitchin 2014c) in which adaptations to urban changes happen immediately, thanks to software able to spot problems as soon as they appear or through bottom-up changes promoted by citizens using widely available portable technologies (Townsend 2013). However, such high expectations of big data arise from the assumption that it can provide information on any kind of subject or urban phenomenon, while in reality the information it provides is incomplete.

The reasons for its incompleteness are at least two. A first reason is due to the limited access that people may have to the technological devices from which big data originates. According to the group taken into consideration, a number of features may reduce the possibility to own and make use of portable devices. For example, an individual with low income may not be able to afford a smart device while someone else may lack the cognitive abilities required to properly make use of it. Such a digital divide questions also the completeness of the information provided by automated and volunteered big data, which only refers to those subjects who are users of portable devices (Graham 2011).

A second element of attention is the fact that, even if an individual has a portable device at her disposal, she may not be mobile. Their non-movement would not be tracked and therefore these subjects are invisible when examining big data. The reasons for not moving can vary. A subject may suffer from a number of impairments (economical, physical and cognitive) or lack usable mobility alternatives and reachable destinations. An individual may be part of relational networks in which subjects are asked not to move and, in exchange, may be able to access goods, services and opportunities without moving (for example, a grandparent may remain at home to take care of their grandchildren while groceries or other relevant goods are bought by other members of the family on their behalf; Plyushteva and Schwanen 2018). A subject may even decide not to move in order to not be tracked, feeling that the ubiquitous tracking technologies leave one with 'no place to hide' (Taylor 2016). These different cases share a lack of movement, which big data is unable to capture. Immobile subjects are therefore not visible, regardless of whether their immobility be voluntary or due to constrained choice.

The incompleteness of the information big data provides questions the possibility to rely on it completely for transport planning. Mainstream approaches are in fact based on a predict-and-provide attitude, in which travel demand is estimated according to visible trends and consequently accommodated with planning and policy interventions (Martens 2006, p. 200). Because of this, "consumers do not escape the constraining have to if they want to enjoy the freedom of having the opportunity to. They have to make a lot of trips in order to be mobile—even in the sense of being potentially able to travel" (Sager 2006, p. 472). Big data thus provides a partial representation of urban phenomena and mobility trends, based on those subjects who are able to move and—voluntarily or not—track themselves while on the move thanks to portable devices. Other groups instead remain invisible to data, so that big data and the algorithms required to process them result in a 'marginalizing power' (Kwan 2016). The incomplete information provided by big data in relation to mobility questions its policy usability, also due to the issues of fairness they raise. Big data emerges thus as a valuable but partial source of information, which cannot be the only knowledge base for understanding and planning everyday mobility.

4.4.2   Data Does Not Speak for Itself

The spread of big data paves the way for a radical epistemological change, transforming the way in which we observe and interpret urban phenomena. The huge amount of available data and the powerful tools to process it may potentially change the approach to scientific research. Instead of facing data to test previous hypotheses, new phenomena and correlations between them emerge as the result of the massive processing of data (Kitchin 2014a). This condition transforms the methodological approaches of empirical sciences that should simply observe what emerges from data and therefore should move from an "hypothetico-deductive method, driven by an incremental process of falsification of previous hypotheses" to "an inductive analysis at a scale never before possible" (Rabari and Storper 2015, p. 33).

Techno-enthusiast approaches go even further and call for 'the end of theory', stating that the quantity of available data makes the scientific method obsolete and instead "calls for an entirely different approach, one that requires us to lose the tether of data as something that can be visualized in its totality. It forces us to view data mathematically first and establish a context for it later. (…) With enough data, the numbers speak for themselves" (Anderson 2008). This move towards an inductive approach not only questions the existence of established disciplines as we know them, but also denies the importance of investigating any aspect that cannot be quantified.

These last excerpts highlight the epistemological risks of simply relying on big data. Considering only this source of information in fact may reduce the possibility to recognize manifold mobility needs and consequently define analyses and actions that contribute to enabling mobilities. At least three risks emerge. First, as mentioned in the previous section, big data necessarily provides an incomplete representation of urban phenomena, grasping only those individuals that can be tracked while also missing the immaterial dimensions that create such phenomena. Second, processing data is not neutral but rather prone to bias and limitations. Despite hopes for an "automatic smart city understanding" (Villanueva et al. 2016, p. 1680), a human-based approach is still required to select and interpret the information deriving from big data. The considerable amount of available unstructured information requires to 'make big data small' (Poorthuis and Zook 2017), selecting only relevant information as well as interpreting what emerging correlations between data are significant or not. The shaping of the tools to examine data is crucial in this sense, considering that non-human agents develop potentially partial ways of understanding the world around them (Mattern 2017) and that some tools, such as algorithms, can act as technical counters to liberty (Greenfield 2017, p. 257). Third, the usability of the knowledge deriving from big data also requires attention. The exclusive faith in the results emerging from data is directly transferred to their use for policy reasons, leading to the belief that "there is one and only one universal and transcendently correct solution to each identified individual or collective human need; that this solution can be arrived at algorithmically (…); and that this solution is something which can be encoded in public policy, again without distortion" (Greenfield 2017, p. 56). This approach nonetheless reinforces a certain 'technological solutionism' (Morozov 2013), according to which "complex social situations can be disassembled into neatly defined problems that can be solved or optimised through computation" (Kitchin 2014b, p. 222).

4.4.3 Not Everyone Is a 'Sensor'

Significant differences emerge even among the subjects who produce data. According to diverse resources, abilities and interests, citizens can act as 'sensors' (Goodchild 2007) in different ways, being simply tracked while on the move or voluntarily providing additional data. In relation to mobility, the data intentionally produced by people can refer to different typologies. First, there are leisure mobilities that are voluntarily tracked by users adding personal information (for example, people recording running or cycling sessions, adding their own physical data as well as elements referring to their running routine). Second, individuals remarking on their mobility experience (for example, people commenting positively or negatively on their commute experience, adding their location, a personal comment and multimedia materials, often as a way to complain about services). Third, geographic information previously missing can be provided, as in the case of open mapping projects based on user-generated information.

Data can stimulate new forms of citizens' engagement and activism. This perspective assumes that individuals have a specific knowledge and command of what data is, thus leading to calls for a specific form of data literacy—that is, "the knowledge of what data are, how they are collected, analysed, visualized and shared, and is the understanding of how data are applied for benefit or detriment, within the cultural context of security and privacy" (Crusoe 2016, p. 12). In relation to community engagement, a specific form of data literacy emerges, definable as "the desire and ability to constructively engage in society through and about data" (Data-Pop Alliance 2015, p. ii). This focus on engagement is reflected also in the short explanations devoted to the terms composing the definition: 1. 'desire and ability' highlights technology as a magnifier of human intent and capacity; 2. 'ability' underlines literacy as a continuum, moving away from the dichotomy of literate and illiterate; 3. 'data' is understood broadly as "individual facts, statistics, or items of information"; 4. "constructively engage in society" suggests an active purpose driving the desire and ability; 5. and 'through or about data' offers the possibility for individuals to engage as "data literate individuals without being able to conduct advanced analytics."(Data-Pop Alliance 2015, p. ii).

Different individuals have different abilities to make use and sense of data. Such differentiation produces relevant imbalances, which question the pervasiveness of data even when considering subjects who are potentially able to produce it. Not everyone is equally able to or interested in acting as a volunteer sensor. More in general, social networks are not fully representative of a society given the different socioeconomic profiles of their users and their different abilities for actively participating in online debates. Consequently, urban phenomena risk being misrepresented. Similarly, volunteered cartographic information may be concentrated in areas where activists are present, neglecting other parts of a territory (Haklay 2013; Neis and Zielstra 2014). In this way, areas that would need more of such user generated information—for example, informal settlements whose cartographic representation in institutional maps is incomplete or imprecise—result in being less represented.

4.4.4 Corporations Are Not Suitable Mobility Planners

Subjects external to public institutions, such as corporations, already develop their own services and tools for urban mobility, heavily relying on small and big data. The growing number of initiatives offering new forms of information or sharing services increases the impact of corporations on urban mobility. The experiences of two private companies, Waze and Citymapper, provide suitable examples of such private contributions. Waze, a GPS navigation software based on a participatory sensing system, collects information from the users and proposes alternative routing to the drivers. In this way, the app contributes to the management of urban traffic, but does so autonomously from the intervention of municipal institutions. Citymapper, instead, is a public transit map and mapping service, whose information is based on user-generated contents, open data and information collected by its employees. In May 2017, the company launched in London a 'smart bus service', a popup service whose routes "show up in A to B routing whenever the algorithm decides so based on their viability and frequency" (Citymapper 2017) and intended to be flexible and open to modifications that accommodate the changes of a city. From an enabling perspective, such initiatives raise concerns related to future forms of mobility management and provision.

Thanks to the data they produce and collect, corporations are getting a more precise, privately owned knowledge of current urban mobility trends. This increasing amount of information determines a higher relevance of private actors when understanding and addressing mobility issues. Despite the praise for the innovative solutions private actors provide, "the critical risk is that this shift in the control of knowledge and associated power will make governing mobility much more difficult in the longer term. The state is already losing its position as the principal source of knowledge about travel patterns on the network relative to mobile phone operators, with this information asymmetry also set to grow further" (Docherty et al. 2017, p. 121). In this respect, the creation of schemes for managing and exchanging the information owned by private and public subjects becomes necessary, although difficult to implement (Vecchio and Tricarico 2019).

Moreover, the information available to corporations also increases their ability to govern urban mobility flows. GPS navigation initiatives like Google Maps and Waze are already the main reference for drivers, given their ability to provide complete information of the real-time conditions of road infrastructures. The search for the most efficient route nonetheless risks going against the mobility plans and strategies of public institutions. For example, it may be that a faster, non-congested route passes right through a neighbourhood where traffic calming interventions have been promoted, leading to an undesired increase in the levels of local traffic. It may even happen that corporations intentionally decide to reroute traffic flows, making it more difficult for people to reach sensitive areas (as may happen in the case of street demonstrations). Even if this seems like a dystopian scenario, the impact that technology may have on the socio-political dynamics of entire countries has been already well demonstrated, for example by Facebook's influence on elections throughout the world (Persily 2017).

Finally, even the provision of services may privilege certain areas and groups over others. In this sense, the impact of sharing mobility services is less clear, as evidence from the United States reveals. For example, a survey showed that ride-hailing services serve minority-majority neighbourhoods (that is, areas where more than half of the residents are racial or ethnic minorities) while taxis usually do not do so (Pew Research Centre 2016). One of the main ridesharing companies, Uber, demonstrated that it well serves areas that are usually underserved, by showing that "median income in a neighbourhood has no meaningful relationship to Uber's level of service in that neighbourhood, including wait times and fulfilment rates" (Uber 2014). Other research instead shows that passengers of these services experience discrimination based on their race and gender, a result that nonetheless depends on the service and on the setting considered (Ge et al. 2016). Similar forms of discrimination may also refer to the scores attributed to the users of these services and may result in differentiated treatment when intending to use both public or private services. This may happen for example when an Uber driver cancels a request due to the low score of a client or, as is increasingly happening in China, where a national reputation system—the Social Credit System—assesses the citizens and eventually excludes them from public services if their score is too low. The pervasiveness of big data configures thus new roles for actors in the mobility field and allows "an unprecedented intervention by a small set of private and unaccountable actors in the structure of opportunity, and the distribution of life chances" (Greenfield 2017, p. 260). The initiatives of these actors raise a number of relevant opportunities but also blind spots when thinking about mobility from an enablement perspective.

4.5   Conclusions

Considering the elements discussed in this chapter, three forms of partiality affect the completeness, the neutrality and the usability of big data in relation to mobility issues. First, the information provided by this newly available data is an incomplete one

(Sect. 4.4.1). Big data offers an unprecedented amount of information, characterized by high levels of detail and by continuous updates, opening new possibilities of knowledge that nonetheless do not grasp some groups and features: the subjects who do not move or do so without being tracked are invisible. Similarly, the reasons behind the mobility or immobility of each subject are also invisible. Second, the tools that make sense of big data are not neutral (Sect. 4.4.2). Algorithms can be constructed differently, highlighting or leaving certain information in the background, despite deriving from the same source. Third, big data can contribute to understanding and planning mobility only if its contribution is considered in relation to the issue to be faced and the specific stage of the policymaking cycle (Sect. 4.4.4). Despite the calls for a data-driven urbanism, in which only data is required to describe urban issues and define relevant solutions, its contribution depends on the use that a number of actors (policymakers, decision makers, technicians and even third-party subjects) may make of them.

The rise of big data and the enthusiasm around its potential for addressing urban issues also raises several problematic features, here considered simply from the perspective of enabling mobilities. Big data can contribute to a fairer mobility planning only if the discussed issues are considered, together with other more generic problems relating for example to privacy and surveillance. Despite the presumably limited amount of individual and collective features that big data can capture, it is possible to imagine that big data will increasingly impact urban planning and policy, as well as the public discourse surrounding them. Because of this, it is significant to increase at least the awareness of what constitutes big data, including its limitations and the specific operational contributions it can provide. In this sense, its integration with other sources of knowledge becomes crucial as it can complement those elements overlooked by big data. Also, the political dimension of big data becomes increasingly important. While this is a non-neutral tool for addressing territorial issues, the number of actors that produce, manage and own data is increasing. These actors are both public and private, with the latter typically being corporations active in fields outside traditional regulations. The rise of big data therefore implies an unprecedented geography of power, configuring a more complex arena in which urban problems are defined, discussed and finally addressed by new constellations of actors. Understanding and exploiting the potential of big data for a fairer urban mobility thus requires equal attention to their technological and political dimensions.

References

Ahas R, Aasa A, Silm S, Tiru M (2010) Daily rhythms of suburban commuters' movements in the Tallinn metropolitan area: case study with mobile positioning data. Transp Res Part C Emerg Technol 18(1):45–54. https://doi.org/10.1016/j.trc.2009.04.011

Amer M, Daim TU, Jetter A (2013) A review of scenario planning. Futures 46:23–40. https://doi.org/10.1016/j.futures.2012.10.003

Anderson C (2008) The end of theory: the data deluge makes the scientific method obsolete. Wired

Banister D, Hickman R (2013) Transport futures: thinking the unthinkable. Transp Policy 29:283–293. https://doi.org/10.1016/J.tranpol.2012.07.005

Bayir MA, Demirbas M, Eagle N (2010) Mobility profiler: a framework for discovering mobility profiles of cell phone users. Pervasive Mob Comput 6(4):435–454. https://doi.org/10.1016/j.pmcj.2010.01.003

Boyd D, Crawford K (2012) Critical questions for big data. Inf Commun Soc 15(5):662–679. https://doi.org/10.1080/1369118x.2012.678878

Brabham DC (2009) Crowdsourcing the public participation process for planning projects. Plann Theory 8(3):242–262. https://doi.org/10.1177/1473095209104824

Chen Z, Schintler LA (2015) Sensitivity of location-sharing services data: evidence from American travel pattern. Transportation 42(4):669–682. https://doi.org/10.1007/s11116-015-9596-z

Citymapper (2017) Say hello to the Citymapper smartbus. Retrieved from https://citymapper.com/smartbus

Cohen N (2018) Algorithms can be a tool for justice—if used the right way. Wired. Retrieved 14 Dec 2018, from https://www.wired.com/story/algorithms-netflix-tool-for-justice/

Crusoe D (2016) Data literacy defined pro populo: to read this article, please provide a little information. J Community Inform 12(3)

Data-Pop Alliance (2015) Beyond data literacy: reinventing community engagement and empowerment in the age of data. Data-Pop Alliance white paper series

Docherty I, Marsden G, Anable J (2017) The governance of smart mobility. Transp Res Part A Policy Pract. https://doi.org/10.1016/j.tra.2017.09.012

Elliott A, Urry J (2010) Mobile lives. Routledge, London

Floridi L (2012) Big data and their epistemological challenge. Philos Technol 25(4):435–437. https://doi.org/10.1007/s13347-012-0093-4

Gandomi A, Haider M (2015) Beyond the hype: big data concepts, methods, and analytics. Int J Inf Manage 35(2):137–144. https://doi.org/10.1016/j.ijinfomgt.2014.10.007

Gartner (2013) Gartner IT glossary—big data. Retrieved 14 Dec 2018, from https://www.gartner.com/it-glossary/big-data/

Ge Y, Knittel CR, MacKenzie D, Zoepf S (2016) Racial and gender discrimination in transportation network companies. National Bureau of Economic Research working paper, 22776. https://doi.org/10.3386/w22776

Gillespie T, Boczkowski PJ, Foot KA (2014) Media technologies: essays on communication, materiality, and society. MIT Press, Cambridge, London

Goodchild MF (2007) Citizens as sensors: the world of volunteered geography. GeoJournal 69(4):211–221. https://doi.org/10.1007/s10708-007-9111-y

Graham M (2011) Time machines and virtual portals: the spatialities of the digital divide. Prog Dev Stud 11(3):211–227. https://doi.org/10.1177/146499341001100303

Greenfield A (2017) Radical technologies: the design of everyday life. Verso, Brooklyn

Haklay M (2013) Citizen science and volunteered geographic information: overview and typology of participation. In: Sui D, Elwood S, Goodchild M (eds) Crowdsourcing geographic knowledge: volunteered geographic information (VGI) in theory and practice. Springer Netherlands, Dordrecht, pp 105–122. https://doi.org/10.1007/978-94-007-4587-2_7

Kitchin R (2014a) Big data, new epistemologies and paradigm shifts. Big Data Soc 1(1). https://doi.org/10.1177/2053951714528481

Kitchin R (2014b) The data revolution. Big data, open data, data infrastructures and their consequences. Sage, London

Kitchin R (2014c) The real-time city? Big data and smart urbanism. GeoJournal 79(1):1–14. https://doi.org/10.1007/s10708-013-9516-8

Kitchin R, Dodge M (2011) Code/space. Software and everyday life. MIT Press, Cambridge, London

Kloeckl K, Senn O, Ratti C (2012) Enabling the real-time city: LIVE Singapore! J Urban Technol 19(10):89–112. https://doi.org/10.1080/10630732.2012.698068

Kwan M-P (2016) Algorithmic geographies: big data, algorithmic uncertainty, and the production of geographic knowledge. Ann Am Assoc Geogr 106(2):274–282. https://doi.org/10.1080/00045608.2015.1117937

Liu Y, Sui Z, Kang C, Gao Y (2014) Uncovering patterns of inter-urban trip and spatial interaction from social media check-in data. PLoS ONE 9(1):e86026. https://doi.org/10.1371/journal.pone.0086026

Lovelace R, Birkin M, Cross P, Clarke M (2016) From big noise to big data: toward the verification of large data sets for understanding regional retail flows. Geogr Anal 48(1):59–81. https://doi.org/10.1111/gean.12081

Lu H, Sun Z, Qu W (2015) Big data-driven based real-time traffic flow state identification and prediction. Discrete Dyn Nat Soc 2015(284906):1–11. https://doi.org/10.1155/2015/284906

Lv Y, Duan Y, Kang W, Li Z, Wang F-Y (2014) Traffic flow prediction with big data: a deep learning approach. IEEE Trans Intell Transp Syst 1–9. https://doi.org/10.1109/tits.2014.2345663

Marsden G, Reardon L (2017) Questions of governance: rethinking the study of transportation policy. Transp Res Part A Policy Pract 101:238–251. https://doi.org/10.1016/j.tra.2017.05.008

Martens K (2006) Basing transport planning on principles of social justice. Berkeley Plann J 19:1–17

Marz N, Warren J (2015) Big data: principles and best practices of scalable real-time data systems. Manning, Greenwich

Mattern S (2017) Mapping's intelligent agents. Places J. https://doi.org/10.22269/170926

Mayer-Schönberger V, Cukier K (2013) Big data: a revolution that will transform how we live, work, and think. Houghton Mifflin Harcourt, Boston, New York

Milne D, Watling D (2019) Big data and understanding change in the context of planning transport systems. J Transp Geogr 76:235–244. https://doi.org/10.1016/j.jtrangeo.2017.11.004

Morozov E (2013) To save everything, click here: the folly of technological solutionism. Public Affairs, New York

Neis P, Zielstra D (2014) Recent developments and future trends in volunteered geographic information research: the case of OpenStreetMap. Future Internet 6(1):76–106. https://doi.org/10.3390/fi6010076

Noulas A, Scellato S, Lambiotte R, Pontil M, Mascolo C (2012) A tale of many cities: universal patterns in human urban mobility. PLoS ONE 7(5):e37027. https://doi.org/10.1371/journal.pone.0037027

Pang B, Lee L (2008) Opinion mining and sentiment analysis. Found Trends Inf Retrieval 2(1–2):1–135. https://doi.org/10.1561/1500000011

Pelletier M-P, Trépanier M, Morency C (2011) Smart card data use in public transit: a literature review. Transp Res Part C Emerg Technol 19(4):557–568. https://doi.org/10.1016/j.trc.2010.12.003

Persily N (2017) The 2016 U.S. election: can democracy survive the internet? J Democr 28(2):63–76. https://doi.org/10.1353/jod.2017.0025

Pew Research Centre (2016) Shared, collaborative and on demand: the new digital economy. Pew Research Centre, Washington

Plyushteva A, Schwanen T (2018) Care-related journeys over the life course: thinking mobility biographies with gender, care and the household. Geoforum 97:131–141. https://doi.org/10.1016/j.geoforum.2018.10.025

Poorthuis A, Zook M (2017) Making big data small: strategies to expand urban and geographical research using social media. J Urban Technol 24(4):115–135. https://doi.org/10.1080/10630732.2017.1335153

Pucci P, Manfredini F, Tagliolato P (2015) Mapping urban practices through mobile phone data. Springer, Berlin

Pucci P, Vecchio G, Concilio G (forthcoming) Big data and urban mobility: a policy making perspective. Transp Res Procedia

Rabari C, Storper M (2015) The digital skin of cities: urban theory and research in the age of the sensored and metered city, ubiquitous computing and big data. Cambridge J Reg Econ Soc 8(1):27–42. https://doi.org/10.1093/cjres/rsu021

Ratti C, Frenchman D, Pulselli RM, Williams S (2006) Mobile landscapes: using location data from cell phones for urban analysis. Environ Plann B Plann Des 33(5):727–748. https://doi.org/10.1068/b32047

Reades J, Calabrese F, Sevtsuk A, Ratti C (2007) Cellular census: explorations in urban data collection. IEEE Pervasive Comput 6(3):30–38. https://doi.org/10.1109/mprv.2007.53

Sager T (2006) Freedom as mobility: implications of the distinction between actual and potential travelling. Mobilities 1(3):465–488. https://doi.org/10.1080/17450100600902420

Schwanen T (2015) Beyond instrument: smartphone app and sustainable mobility. Eur J Transp Infrastruct Res 15(4):675–690

Sevtsuk A, Ratti C (2010) Does urban mobility have a daily routine? Learning from the aggregate data of mobile networks. J Urban Technol 17(1):41–60. https://doi.org/10.1080/10630731003597322

Soto V, Frías-Martínez E (2011) Automated land use identification using cell-phone records. In: Proceedings of the 3rd ACM international workshop on MobiArch

Taylor L (2016) No place to hide? The ethics and analytics of tracking mobility using mobile phone data. Environ Plann D Soc Space 34(2):319–336. https://doi.org/10.1177/0263775815608851

Townsend AM (2013) Smart cities: big data, civic hackers, and the quest for a new utopia. W.W. Norton and Company, New York, London

Uber (2014) Uber economic study: Uber serves underserved neighborhoods in Chicago as well as the Loop. Does taxi? Retrieved 17 Dec 2018, from https://www.uber.com/blog/chicago/uber-economic-study-uber-serves-underserved-neighborhoods-in-chicago-as-well-as-the-loop-does-taxi/

Vecchio G, Tricarico L (2019) "May the force move you": roles and actors of information sharing devices in urban mobility. Cities 88:261–268. https://doi.org/10.1016/j.cities.2018.11.007

Villanueva FJ, Aguirre C, Rubio A, Villa D, Santofimia MJ, López JC (2016) Data stream visualization framework for smart cities. Soft Comput 20(5):1671–1681. https://doi.org/10.1007/s00500-015-1829-8

Zikopoulos P, Eaton C (2011) Understanding big data: analytics for enterprise class hadoop and streaming data. McGraw-Hill, New York

Chapter 5
Stations: Nodes and Places of Everyday Life

Abstract This chapter introduces the role of medium-small railway stations within daily mobilities and the conditions that enhance their role in large metropolitan areas characterized by a dispersed demand and mostly oriented towards the use of cars. With the aim of proposing tools for coordinating public accessibility and land uses, the chapter highlights how strengthening the regional railway supply, as done in some Italian regions, could represent also a land-use policy for re-orienting urban settlements and land-use forecast. Starting from a reflection on the outcomes of an investment in upgrading the regional railway service in the Lombardy Region, the chapter proposes a classification of the stations as a useful tool for the construction of scenarios that reorganize land-use forecasts and improve both the accessibility and quality of the services in the stations, in order to widen the catchment areas of each station. The approach combines two methodologies of classification of the stations, both able to enhance the place and node dimensions of each station: the 'Place-Node model' (Bertolini 1999) and the 'TOD index approach' (Evans and Pratt 2007). The classification of the 104 suburban railway stations provides guidelines for densification around some stations, reorganizing dispersed land-use forecast and improving the quality of the transport connectivity and the railway services.

Keywords Railway stations · Classification · Lombardy Region · Node-Place model · TOD index

5.1 Introduction

The station is a node where interconnected transport and technical networks offer differing degrees of accessibility to other places and, at the same time, is a place belonging to an urban/peri-urban context. It is characterized by different users and uses, different densities and rhythms, related not only to the transport supply, but also to the permanently or temporarily inhabited areas around it.

This chapter was authored by Paola Pucci.

Due to these features, the station belongs to a transport network and simultaneously to a spatial setting, that is, the place where the accessibility offered by transport becomes spatial capital. The station provides the accessibility that offers individuals the opportunities of performing activities and participating in social life. Therefore, the station mobilises spatial capital (Pucci 2019) and, based on the connectivity it offers, it is 'a condensation' of various forms of spatial capital.

This interpretation is informed by three approaches that, in different ways, have investigated the role of the station in the 'process of territorialisation' of the transport networks, with respect to the mobility practices and uses. It also supports policies aimed at coordinating transport accessibility and land uses.

The first approach is based on a re-conceptualisation of the infrastructure network as a 'territorial network' (Dupuy 1991), which involves enhancing the role of the nodes as the main 'objects' where the network meets the territory, with important consequences in addressing process of territorial organisations. The network, as a "dynamic tool for structuring urban settlements" (Amar et al. 1991, p. 7), through the role of the nodes—as a main interconnection points—becomes porous, open to and into space, overcoming an idea of the transport network as a closed and tubular system, regulated only in terms of flows. In this interpretation, the performances of the network depend on the quality of the interconnections between and into the nodes. It concerns the technical interconnections that determine the level of accessibility offered; the local interconnections that deal with the urban/peri-urban context in which the station is located and its connectivity, through which the catchment area of each node is defined, and finally the internal interconnections characterized by the distributions of spaces into the nodes, affecting the organisations and the equipment of the spaces and paths, as well as the management dimensions (Amar et al. 1991).

A second approach, focusing on the practices of use of the stations, treats them as a place of everyday life, as a *tiers lieu* between home-place and work-place. In doing so, the station is interpreted as a social microcosm or as 'a theatre of experimentation', with new uses and activities, involving different modes of sociability (Tillous 2009, 2016), thus bringing new insights into research led by Isaac Joseph (1995) and by Michel Kokoreff (2002). Through these interpretations, a new image of the relationship between station and urban/peri-urban contexts is promoted and which involves also the promotion of new services and activities (synthesizable in some formats as 'Pickup Station', 'Market and Station' and 'Work and Station') that participate in the construction of urban opportunities, especially in peri-urban contexts characterized by a marked mono-functionality.

The third approach investigates the role of the transit node within the 'land-use transport feedback cycle', for supporting policies finalised to coordinate transport accessibility and land uses.

Exploring more in depth the ways in which the 'land-use transport feedback cycle' has been implemented in policy focused on transportation and land-use integration, the Chapter introduces the main approaches using accessibility to transit stations as a key condition for reducing the negative externalities of economic and land-use growth, as well as for enabling mobilities (Sect. 5.2). Selecting two approaches to classify railway stations, Sect. 5.3 develops a classification of the suburban railway stations in the Lombardy Region (Northern Italy), applying both the Node-Place model (Bertolini 1999) and the TOD index approach (Evans and Pratt 2007). Comparing the final outcomes of both approaches (Sect. 5.3.3), the Chapter aims at providing a general framework for addressing more sustainable land-use and transport policies and to evaluate an infrastructural investment in railway supply (Sect. 5.4).

5.2   Coordinating Transport Accessibility and Land Uses: The Role of the Stations

A well-established literature (Wegener and Fürst 1999; Meyer and Miller 2001) has conceptualised the complex two-way dynamic connection between land-use and transportation systems through the 'land-use transport feedback cycle' where land use related factors and transport patterns both affect each other in a mutual way (Fig. 5.1).

Despite this model simplifying the complexity of the effects of integrated land-use transport policies, as they depend on a multitude of simultaneous changes of related system variables[1] and the practice of transport planning is far more complex and

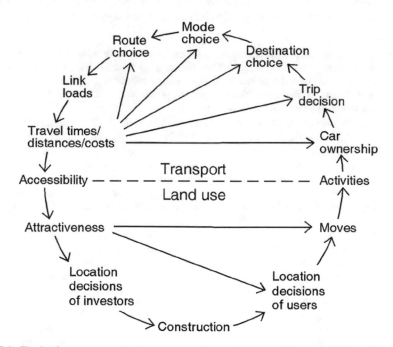

Fig. 5.1 The land use-transport feedback cycle. *Source* Lautso and Wegener (2007, p. 154)

[1]As argued by some authors (Van Wee 2002; Bertolini 2012), under real conditions, land use development depends on a multiplicity of factors, but is also of a conjectural nature and difficult to

'wicked' (Rittel and Webber 1973), the 'land-use transport feedback cycle' is a valid conceptual framework for supporting policies intending to add value to the transport nodes. The 'land-use transport feedback cycle', addressing integrated policies of land use and transport, emphasizes the potentialities of transport nodes, based on their level of accessibility that create opportunities for spatial interaction.

Some well-established policy and programs—like TOD[2] approaches, ABC[3] and Vinex[4] programs or the Bahn Ville project[5]—address the issue of coordinating land use and travel patterns, using accessibility to transit stations as a key condition for reducing the negative externalities of economic and land-use growth.

Among them, Transit Oriented Development (TOD) is one of the most relevant "viable model for transportation and land-use integration in many developed and rapidly developing cities of the world" (Cervero 2009, p. 23). As a very important part of a broader smart growth approach to urban development, TOD pursues a combination of transit, walking and cycling environments that consist of forms of walkable, mixed-use urban developments surrounding new and existing rail or rapid bus stations, planned and designed, in most cases, as part of a larger regional network. Tested in the US and in Canada as an integrated policy combining regional planning, city revitalization, suburban renewal, walkable neighborhoods and conceived within the context of at least a corridor or a regional metropolis, the TOD approach has spread widely in Asia and Australia as being consistent with the 'smart growth' movement. It has become a kind of urban design archetype in the integration of

---

predict (e.g. trends of the regional demand and real estate market, availability of land, attractiveness of the local environment and in general land use policies). At the same time, the development of transport systems is not only determined by the demand for movements generated from land use, but also by infrastructure investments and progresses on the supply side, such as technological innovation or transport policy (Bertolini 2012, p. 19).

[2] The word TOD was first introduced by Peter Calthorpe in his book *The Next American Metropolis: Ecology, Community, and the American Dream,* 1993, Princeton Architectural Press, New York.

[3] Promoted by the Ministry of Housing, Planning and Environment (1990) in the Netherlands, ABC policy is based on the purpose of concentrating activities with a high mobility demand in places with good levels of accessibility of public transport, accompanying this option with a policy of restriction in car parking areas. ABC policy defines a strategy of localization for activities (business and services) crossing the 'profile of accessibility' of a place with the 'mobility profile' of an activity (in terms of people and goods attracted).

[4] The VINEX Program within the Fourth Report on Spatial Development (Extra) of the Netherlands intervenes on population density (standard plan 34.3 houses/ha) and on choice of location for new residential expansions in which the issue of accessibility to public transport takes the central role. The program identifies three different situations: *infilling locations*, where the new building affects built fabrics and areas near urban centers (distance 5 km, average distance home/work less than 10 km) and public transport nodes; *expansion locations*, covering an area outside a distance of 5–10 km from the city center and between 10 and 15 km from the workplace; *outer areas*, or outside areas located near an existing train station or a one in construction (more than 25 km from the workplace and more than 10 km from the nearest urban center). VINEX agreements came in the form of package deals, comprising subsidies for land development, soil sanitation, regional recreational areas, and urban and regional public transport and roads, plus some extra stops in the national train system.

[5] A French-German action research project finalized to develop successful strategies for integrating land-use and transport around attractive regional railways.

a certain degree of density, diversity and pedestrian-oriented designs, built around existing and future public transport stations.

Described by some authors as an 'emerging European-style planning in the USA' (Renne and Wells 2004), TOD planning principles are evidently traceable in different European experiences (Pojani and Stead 2016; Staricco and Vitale Brovarone 2018): from the Garden Cities by Ebenezer Howard in the late 1800s, and the New Towns built in Europe in the mid-20th century (Renne 2009a, b) to the Finger Plan of Copenhagen (Knowles 2012) and the 'Planetary Cluster Plan' of Stockholm (Cervero 1995). The *ciudad lineal* by Soria y Mata planned the urban development in 1886 as a linear corridor in association with the electric tram, which looks similar to the Transit boulevard by Calthorpe (2001), almost in a didascalic way, as well as in the coordination between the development of streetcar (electric tram), underground and commuter railway routes and a star-shaped urban form that emerged in the late 19th and early 20th centuries in the US, financed at least in part by land value capture (Knowles 2006).

The literature about TOD experiences and their practical applications have produced numerous reflections on the necessary conditions for the implementation of this approach (in terms of location efficiency, mix of choices, value capture and place making), as well as on the barriers that can prevent the realisation of TOD principles (Curtis 2008; Curtis and Low 2012; Filion and McSpurren 2007; Haywood 2005; Curtis et al. 2009). Indeed, because 'TOD is not an island', the strategies for establishing TOD as a pattern of urban development deal with transport and land use integrations, as well as with governance challenges. Newman (2009, p. 13) recognizes four strategic planning tools for TOD: "a strategic policy framework that asserts where centres need to occur and at what kind of density and mix; a strategic policy framework that links centres with a rapid transit base, almost invariably electric rail; a statutory planning base that requires development to occur at the necessary density and design in each centre, preferably facilitated by a specialized development agency, and a public-private funding mechanism that enables the transit and the TOD to be built or refurbished through a linkage between the transit and the centres it will service" (p. 13).

Despite many of the arguments for pursuing TOD being similar, TOD principles cannot be applied equally in different contexts, nor even in the same city or metropolitan region. Particular land use features may surround each individual station and transport supply determines the multimodal accessibility at each transit node. These are relevant conditions for addressing the density and diversity of functions around each transport node.

As argued by Kamruzzaman et al. (2014), the question is not whether a site is suitable or not, but rather for what type of TOD (if any) or not. Based on this, the challenge is "to develop a general typology of places to account for a variety of different scales (large city, small city, town), locations in the metropolitan area (central city, peripheral city, commuter town), transit type (commuter rail, frequent light rail), and other key attributes" (Belzer and Autler 2002, p. 30).

The need for developing typologies for TOD has recently been the object of several experimentations and empirical applications as few studies have empirically generated TOD typologies or used performance indicators in a quantitative way. In this perspective, Kamruzzaman et al. (2014) and Lyu et al. (2016) reconstruct the approaches aimed at categorising TOD into typologies as a relevant tool "for enhancing their planning, design, and operational activities and supporting the identification of general development potentials and necessary future adaptations" (Kamruzzaman et al. 2014, p. 55).

From this overview of the recent findings, different classifications can be sorted in different ways, mainly identifying comparable stations with respect to several combinations of transport and settlement features (node and place types). This task can be undertaken in order to promote collective actions, aimed at reducing the management complexity for infrastructure companies (application of standards in operation and development), securing the consistency of actions across large portfolios and geographic regions or for identifying the sites and actors with comparable challenges and experiences. At the same time, classification also helps in identifying successful benchmarks, highlighting the need for actions to be considered as best practices replicable in other cases and in supporting the identification of general development potentials and necessary future adaptations of whole classes and within classes of stations (Zemp et al. 2011, p. 670).

Among the methods of developing the TOD typology, there are two main approaches able to deal with the potentialities of the transport node by operationalizing the 'land-use transport feedback cycle'. The first is the conceptual framework of a 'Node-Place model' developed by Bertolini (1999) and its subsequent applications in Reusser et al. (2008), Zemp et al. (2011), Vale (2015, 2018) and more recently in Nigro et al. (2019). The second approach can be found in the 'TOD Index' proposed by Evans and Pratt (2007), as well as by the Center for Transit-Oriented Development (2010) that has taken into account both a place indicators (e.g. use-mix) and a performance indicators (e.g. household VMT) to develop TOD typologies in the USA.

In both approaches, the main criteria for determining a TOD typology are related to the 'double feature' of the transport junction as a node characterized by different forms of accessibility and as a place belonging to an urban/peri-urban context where different uses take place.

5.3   Classifying Railway Stations in the Lombardy Region

The research proposes a classification of railway stations as a useful tool for scenario construction, aimed at reorganizing land use forecasts and improving the quality of services and connectivity in the suburban railway stations in Lombardy (Northern Italy) to widen their catchment areas and potential train users.

The experimented approach applies, in a comparative way, two methodologies of classification of the stations, both capable of enhancing the dimension of 'place' and of 'node' related to each transit station: the Node-Place model (Bertolini 1999) and the TOD index approach (Evans and Pratt 2007).

The application of both methods aims at verifying their effectiveness in identifying stations characterized by problems and/or opportunities in order to lead to
precise policy actions. The goal is to either improve services, densifying the areas
around some stations or to transfer build-volumes from areas without public transport
accessibility.

The study involved 104 railway stations of the Suburban Railway Network (S
lines) of the Lombardy Region (Fig. 5.2), widely distributed across a territory where
39% of the inhabitants live less than 1 km from a suburban railway station.

The upgrading of the suburban rail service in Lombardy, and in particular in the
Milan Urban Region, started in 2002 with the improvement of the supply and the
overall management of the railway lines, which were arranged into recognisable
services running at a high frequency (at least every 30 min, every day of the year).
Such enhanced service introduced new challenges for the integration between urban
settlements, land use forecasts and railway accessibility, in particular in low-density
territories where transport demand is more dispersed and car-oriented (Pucci 2017).

Since 2002, the suburban railway supply has been improved in the number of daily
connections with an increase in the rail supply estimated by the Lombardy Region,
to the order of more than 50% (Regione Lombardia 2016). This has been achieved
through a scheduled time of two trains per line each hour per direction, thanks to the
doubling and securing of the tracks and the construction of new stations.

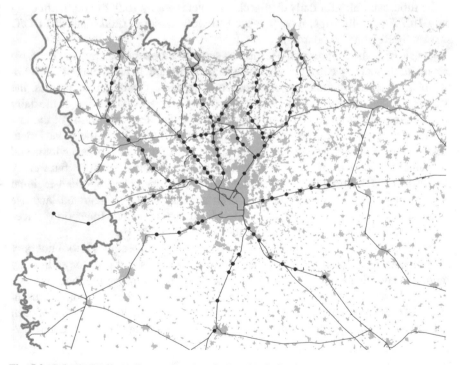

Fig. 5.2 Suburban railway lines and stations in Lombardy Region

Table 5.1 Distribution of the demand by type of railway service and loading index (2018)

	Express regional lines	Regional lines	Suburban lines	total
Lines	10	36	13	59
Trains/day	415	110	936	1461
Passengers/day	165.000	224.000	311.000	700.000
%	23.57	32.00	44.43	100.00
Passengers/train	397.59	2036.36	332.26	479.12

In Table 5.1 the distribution of the demand by type of railway service shows the role of the suburban lines as the most used railway service among the railway system, with a percentage of 44.4% of passengers per day, but an average of passengers/trains lower in respect to the regional lines. This is due to the features of the suburban service, in particular its capacity to carry high flows over short distances and, at the same time, by serving the territory in a capillary way, thus guaranteeing a high frequency of trains.

Despite the important public investments for improving the railway services, the modal share in favour of the railway lines did not change significantly. The use trends of the suburban trains for daily displacements increased by 10.9% overall, affecting only 10.7% of daily trips, mostly related to work-reasons[6] (2002 and 2014 O/D surveys by Regione Lombardia).

Considering the features of the territories served by the suburban railway lines, we can highlight that in the low density urban settlements and, in particular, in the South of Milan where the stations are often eccentric in respect to the urban centres, the use of the train is lower also for work-related displacements (about 6% of the daily commuters displacements). This little aggregated data effectively warns of the low impact of the public investment in the railways on mobility habits in the Milan Urban Region. The reasons can be sought not only in terms of the mobility behaviours and individual preferences of those who live, work, travel in this territory, but mostly in terms of the sectoriality of the implemented measures. The important regional investment in the rail services has not been accompanied by integrated land-use policies, as well as by coordinated measures for reorganizing bus-transport services to affect mobility practices in favour of the train (Pucci 2015b).

In this framework, the classification of railway stations becomes a tool not only for categorizing different performances of railway stations, but also for addressing actions to coordinate public accessibility and land uses, improve the local accessibility to the stations to enlarge their catchment areas and affect the modal share in favour of the train. The classification of the 104 stations evaluates the profile of each of them by setting up actions aimed at enhancing their intermodality and connectiv-

---

[6]Considering the modal share of the daily displacements, 5.1% of the work-related movements are made by train, a percentage that reduces to 3.5% for occasional trips (Regione Lombardia 2014).

ity, widening their area of influence, selectively densifying the areas around some of them as well as reorganizing the land use of forecasted sprawl and car dependency.

In this approach, accessibility is not just a feature of transportation. Rather, it is the quality and quantity of transport supply offered in the node (transport accessibility) that defines the role of the railway station at the macro scale. At the same time, the quality of the pedestrian and bike connectivity to the station and the quality of activities and services (local accessibility) that define its local catchment area. Through a cross-scale approach, the station becomes a place where different scales are combined to address urban and transport policy.

To reach these goals, the Place-Node model and TOD index were applied at the 104 railway stations of the Suburban Railway Network of the Lombardy Region. This was done by considering two buffers of 400 and 1500 m around each station to define its catchment area, with respect to walking (400 m) and to the accessibility with motorized vehicles and/or by bicycle (1500 m).

These buffers differ from other experiences where a single buffer of 700 m was used (Bertolini 1999; Reusser et al. 2008; Zemp et al. 2011, Vale 2015; Chorus and Bertolini 2011). As argued by Reusser et al. (2008), if "a 700 m distance was defined by Bertolini (1999) as appropriate buffer for densely populated areas such as the Netherlands and Switzerland, other distances may be considered for other countries, but further research is required on catchment size of railway stations. Using other distances has influence on the indicators for few stations, but does neither change the general relationship between node and place index nor the interpretation of the clusters" (Reusser et al. 2008, p. 195).

5.3.1 Node Place Model Application

The Node-Place model (Bertolini 1999) offers a conceptual framework to address a public transport-oriented development that focuses on two constituent dimensions of the stations and defined by the interaction of the node and place value.

The node value expresses the accessibility of the node and thus "its potential for physical human interaction" (Bertolini 1999, p. 201). The place value represents the intensity and diversity of activities, and thus "the degree of actual realisation of the potential for physical human interaction" (Bertolini 1999, p. 201).

The Node-Place model distinguishes five ideal-typical situations for a station area, in which every regional node takes its position on the node-place scale, reflecting its condition in the hierarchy of the regional system (Fig. 5.3).

This is a dynamic model because the position of each station can evolve based on the criteria of the 'land-use transport feedback cycle'. The transport provision (the node value) of a location can increase by improving accessibility, which creates the favourable conditions to the further reaches of the location. In turn, the development of a location (the place value), because of a growing demand for transport, creates favourable conditions for the further development of the transport system (Bertolini 2012).

Fig. 5.3 Node-Place model
by Bertolini. *Source*
Bertolini (1999, p. 202)

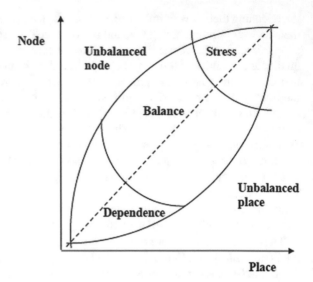

Based on this, the Node-Place model is an effective tool to identify opportunities for intensification and differentiation of urban activities around public transportation nodes or to recognize those situations in which a reorganization of the public transport network is needed to enhance the accessibility of disconnected places.

The model establishes a new planning method able to handle the complexity of transport multimodality at a regional level. Even if it cannot predict development, it can be used to gain a better understanding of development dynamics and discover locations suitable for development in a metropolitan or regional area.

Following this condition, it is expected that the unbalanced situations tend to evolve into a more balanced state, which might happen in three different ways (Reusser et al. 2008). An unbalanced node might increase its place index, for instance by attracting urban development, it might decrease its node index, for instance by reducing the transportation supply, or pursue both strategies at the same time. An unbalanced place might increase its node index, decrease its place index (probably more difficult and unlikely) or pursue both at the same time. In any case, the Node-Place model can be used as an analytical framework to guide urban and transportation planning to promote more balanced, integrated and transit-oriented sites (Bertolini 1999; Chorus and Bertolini 2011).

The Node-Place model has been experienced and used, since its elaboration, in different contexts and ways (Nigro et al. 2019), which have contributed to elaborating an 'extended version' of this approach, in terms of parameters considered and processed (Reusser et al. 2008; Zemp et al. 2011; Caset et al. 2018; Vale et al. 2018; Lyu et al. 2016). In particular, the main additional contributions refer to considering different modes of access to stations and therefore adopting various radiuses for defining the catchment area of each station (Caset et al. 2018) and for introducing a 'design index' to investigate the factors that influence the pedestrian accessibility of stations' catchment areas (Vale et al. 2018).

This research applied Node-Place model selecting indicators capable of identifying 'node and place' profiles, considering both the parameters processed in previous research, as well as the availability of data (Table 5.2).

Table 5.2 Node and place indicators

	Indicators	Used in our research	Studies which process it
Node index	N. of directions served by train		Bertolini (1999), Reusser et al. (2008), Chorus et al (2011), Vale (2015), Vale et al. (2018)
	Daily frequency of railway supply		Bertolini (1999), Reusser et al. (2008), Zemp et al. (2011), Vale (2015), Vale et al. (2018)
	N. of stations with 45 minutes of travel		Bertolini (1999)
	N. of stations with 20 minutes of travel		Reusser et al. (2008), Zemp et al. (2011), Vale (2015), Vale et al. (2018)
	Type of train connection		Chorus et al. (2011)
	Daily passengers get in/get off		
	N. of directions served by LPT		Bertolini (1999), Reusser et al. (2008), Chorus et al (2011), Vale (2015), Vale et al. (2018)
	Daily frequency of TPL supply		Bertolini (1999), Reusser et al. (2008), Zemp et al. (2011), Vale (2015), Vale et al. (2018)
	Distance from he closet motorway access		Bertolini (1999), Vale (2015), Vale et al. (2018)
	Car parking capacity		Bertolini (1999), Vale (2015), Vale et al. (2018)
	N. of free-standing bicycle paths		Bertolini (1999)
	Bike path length within 2 km		Reusser et al. (2008)
	Bicycle parking capacity		Bertolini (1999)
Place index	N. of residents		Bertolini (1999), Reusser et al. (2008), Chorus et al. (2011) Zemp et al. (2011), Vale (2015), Vale et al. (2018)
	N. of residents' labour force		Bertolini (1999), Reusser et al. (2008), Chorus et al. (2011), Vale (2015), Vale et al. (2018)
	N. of residents' students		Bertolini (1999), Reusser et al. (2008), Chorus et al. (2011), Vale (2015), Vale et al. (2018)
	N. of homes occupied by residents		Bertolini (1999), Reusser et al. (2008), Chorus et al. (2011), Vale (2015), Vale et al. (2018)
	N. of workers in retail/htel/catering		Bertolini (1999), Reusser et al. (2008), Chorus et al. (2011), Vale (2015), Vale et al. (2018)
	N. of workers in education/health/culture		Bertolini (1999), Reusser et al. (2008), Chorus et al. (2011), Vale (2015), Vale et al. (2018)
	N. of workers in industry and distribution		Bertolini (1999), Reusser et al. (2008), Chorus et al. (2011), Vale (2015), Vale et al. (2018)
	N. of workers in administration&services		Bertolini (1999), Reusser et al. (2008), Chorus et al. (2011), Vale (2015), Vale et al. (2018)
	N. of jobs		Zemp et al. (2011)
	Surface of industrial activities		
	Surface of commercial&tertiary activities		
	Surface of services		
	Degree of functional mix		Bertolini (1999), Reusser et al. (2008), Chorus et al. (2011), Vale (2015), Vale et al. (2018)

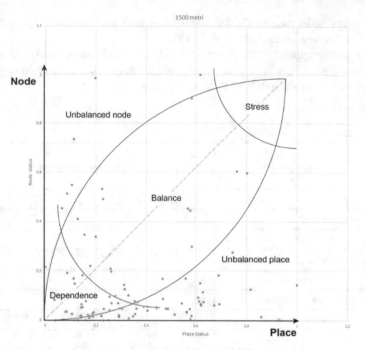

Fig. 5.4 Classification of the suburban railway stations with Node-Place model: buffer of 1500 m

The approach allows to obtain for each station a Zscore[7] that relates to the transport features (*node status*), as well as to the density and diversity of the urban settlements (*place status*), divided into two buffers around the stations (Fig. 5.4).

Thanks to a correlation analysis of a set of indicators that describe the node and place dimensions (Zemp et al. 2011), the approach identifies the relative position of each station, recognizing five ideal-typical situations described in Fig. 5.5 and summarized as follows:

- The unbalanced places are concentrated in the densest urbanized area in the North of Milan, mixed with balanced situations. Few other unbalanced places are localised in those areas at the South of Milan, recently affected by an intense urban growth.

---

[7]For each indicator a Z-score has been calculated, following this formula:

$$z = \frac{x - \mu}{\sigma}$$

where
  x is the normalized indicator;
  μ is the mean of the indicator;
  σ is the standard deviation of the indicator.

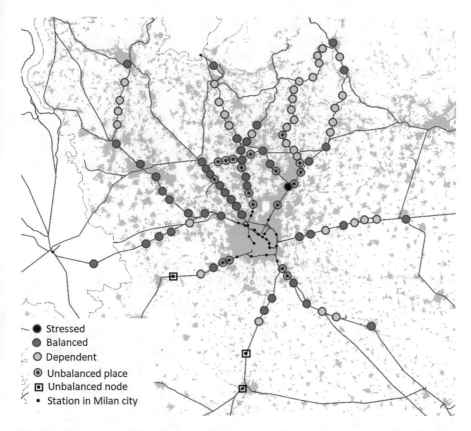

Stressed
Balanced
Dependent
Unbalanced place
Unbalanced node
Station in Milan city

Fig. 5.5 Classification of the suburban railway stations with Node-Place model (buffer of 1500 m)

- Dependent railway stations with a low level of node and place dimensions are located in the Northern foothills area along the railway lines to the regional capitals of Varese, Como and Lecco.
- The unbalanced nodes emerge only in the buffer of 1500 m and they are localised in the low-density areas in the South of the Milan Urban Region.

The balanced situations are the most recurrent conditions among the railway stations (44% of the suburban stations), followed by the dependent situations (35% of the suburban stations).

5.3.2 TOD Index Application

The TOD index measures the 'TOD-ness' of a place (Singh et al. 2017) by considering various indicators that characterise a TOD, in order to identify those areas with characteristics able to stimulate the development of a Transit Oriented strategy. Despite Renne and Wells (2004) arguing that there is a lack of comprehensive tools to quantify the existing levels of TOD and identify areas where TOD can be developed, Evans and Pratt (2007) describe a TOD Index as a measure of "potential device for considering the degree to which a particular project is intrinsically oriented towards transit" (Evans and Pratt 2007, p. 17–3).

Therefore, quantifying TOD in terms of an index implies computing and combining the factors that influence a Transit Oriented Development (TOD): compactness and density in the built-up area; land-use and social mix; high accessibility by public transport and intermodality or a liveable urban environment with a high quality of public spaces and pedestrian paths.

In order to combine these features into a comprehensive TOD index, a comparative analysis has been carried out on the main operative experiences (Renne 2009a, 2009b; Singh 2015; Singh et al. 2017; Ngo 2012) to find the most suitable indicators for measuring TOD levels (Table 5.3).

The selected indicators have been processed using spatial analytical tools (GIS) and Spatial Multiple Criteria Analysis (SMCA) to calculate a TOD Index, identifying

Table 5.3   Selected indicators for the potential TOD

	Criteria	Rank	Weight	Indicators	Contribution to criterion %	Associated variables
Built environment	Potential transformability	2	0.24	Brownfield	40	Sqm of brownfield / Buffer area
				Transformation areas in Urban Plan	10	Sqm transformation areas / Buffer area
				Vacant areas	50	Sqm vacant areas / Buffer area
	Densification	3	0.19	Density of buildings	50	Residential buildings < 3 floors/total residential buildings
				Density of inhabitants	50	Low residential population / Buffer area
Travel behaviour	Promote sustainable travel behaviour	4	0.14	Car dependency	75	Number of workers and students that use car to move (within 15 min)
				Suburban distance of commuting	25	Number of workers and students with a trip's time not higher than 30 min (within 15 min)
	Valorise transit node	1	0.29	Low use of transit capacity	100	Low utilisation density of service: number of transit's users/frequency
Local economy	Valorise existing facilities	6	0.05	Attractive services	100	Attractive services/Number of trains at day
Natural environment	Protect environment	5	0.1	Presence of protected areas	100	Sqm natural protected areas/Buffer area

existing and potential Transit Oriented sites[8]. Through a multi-criteria analysis[9] that allowed different weights to be assigned to the selected criteria, it was possible to obtain for each station a score that expresses the level of propensity of each station to be a TOD and therefore to be the object of densification policies.

This analysis was possible thanks to a distinction made between the existing conditions characterizing the station as a TOD (existing TOD) and potential conditions (potential TOD). This difference was expressed in particular by the availability of transformation areas around the node, making the station a potential place to be densified.

The Potential TOD Index allows the identification of sites around transit nodes, which have the characteristics to implement a TOD strategy. Based on this, the Potential TOD Index measures the Potential TOD levels in areas within walkable limits of the suburban railway stations, permitting the implementation of specific strategies for each station according to the weight of each indicator. Starting from the ideal set of indicators in the studies, the selected list of indicators and the relative weights used in this research are summarized in Table 5.3.

The outcomes of this approach highlight Potential TOD sites as widely distributed within the whole Milan Urban Region (Fig. 5.6), allowing an identification where stations with TOD conditions can be improved. Potential TODs are characterised by their high level of potential transformability and environmental preservation. In other words, there exists the possibility to densify the area around the railway station, avoiding soil consumption and negative environmental impacts. In addition, they are characterized by a low density, as well as a low level of saturation of the transit capacity. This condition is very important since an increase in the intensity of inhabitants, jobs and services will translate to an increase in the railway service's users.

---

[8]The analysis has been carried out by Alessio Praticò in the Master Degree Thesis (2017) "Coordinating accessibility and land use. Tools and strategies to classify railway stations in the Milano Urban Region, Politecnico di Milano, supervisor Paola Pucci.

[9]The formula is as follow:

$$TOD\,index = \sum_{j=1}^{n} w_j a_{ij}$$

where

    w_j = is the relative weight attributed to the criterion C_j
    a_{ij} = is the value of the alternative A_i based on the criterion C_j

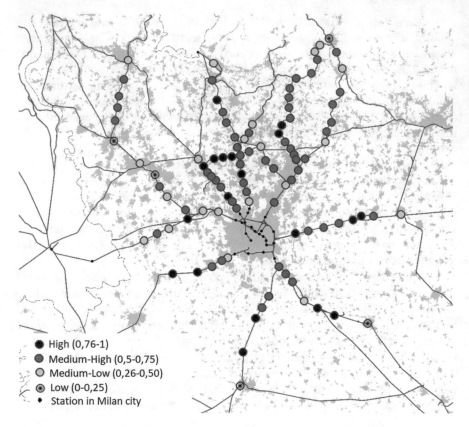

Fig. 5.6 Potential TOD classification

5.3.3   *Comparing Node-Place Model and Potential TOD Index*

The experimentation of two methods in the same territory makes it possible to evaluate their effectiveness in terms of the results achieved, their ability to address policy measures for improving the role of the stations, as well as in terms of reproducibility.

With respect to the results and to their ability to address policy measures, both the Node-Place Model and the TOD Index offer wide ranging possibilities to compare different cases, in order to identify similar patterns and consequently similar strategies of intervention.

The overlapping of the final outcomes of both approaches shows a partial coincidence between unbalanced stations of the Node-Place model and Potential TOD. In the most urbanized areas (Northern Milan Urban Region), the high presence of 'Unbalanced places' and 'Potential TOD' suggests actions focussed more on increasing the accessibility and transport supply of the stations. Instead, some stations in

Southern Milan are characterized as unbalanced nodes and places in which to propose densification policies.

In terms of reproducibility, the visualization of the outputs of Node-Place model within the Cartesian diagram allows a quick identification of the main profile of each railway station, so as to associate possible policy measures for structuring station area redevelopments.

On the other hand, the construction of the TOD index is more complex and involves a multi-criteria analysis that processes a more extensive number of qualitative and quantitative indicators. In doing so, despite being more complex to reproduce, it is also more punctual in defining the typologies of the stations for addressing precise policy actions.

5.4 Policy Implications: Stations of Everyday Life

The results obtained from the two classifications tested on the 104 suburban railway stations allow for the selection of the stations in need of densification policies, as they are characterized by a good public transport supply, low urban densities and/or the availability of transformation areas ('Unbalance node' and 'Potential TOD'). At the same time, there are stations with high transport demand and significant density of displacements not supported by an adequate local public transport supply and mobility services ('Unbalanced place').

In particular, the situations that promote policy aimed at enhancing railway upgrading are: stations with a high 'Potential TOD index' values and stations belonging to the 'Unbalanced places' and 'Unbalanced nodes' types (Fig. 5.7).

In the 'Unbalanced places', we can find high densities of land uses where it would be profitable to address actions towards the improvement of accessibility to the stations, with the aim of widening the potential basin of train users. In particular, it is a matter of enhancing the quality and efficiency of the connectivity paths through public transport and cycle-pedestrian links, but also by encouraging car pooling, car sharing and equipping spaces and services devoted to different forms of shared mobility.

The 'Potential TOD' and 'Unbalanced nodes' are stations with excellent performances in terms of transport supply and accessibility and could accommodate new volumes, as they are adjacent to abandoned or unused areas, available for densification projects.

In these stations, the densification processes must be commensurate with the transport capacity of the node, in order to avoid a malfunctioning of the transport supply. Above all, they must be the result of regional compensation policies devoted to coordinating land use and transport planning that will prevent competition between municipalities for tax paying residents and firms and conflicting interests among different stakeholders within the cities.

Possible scenarios of land use densification have to be defined from a re-organisation of the local land use forecasts in municipalities not served by an adequate

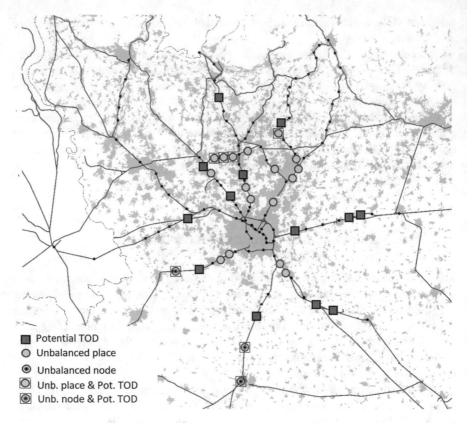

■ Potential TOD
○ Unbalanced place
◉ Unbalanced node
◻ Unb. place & Pot. TOD
◩ Unb. node & Pot. TOD

Fig. 5.7 Railway stations where to address mobility and land use policy

public accessibility, thus avoiding new soil consumption and assuming volumetric transfers, based on shared criteria to preserve the integrity of non-urbanized territory (Pucci 2015a, b).

In both situations, actions aimed at increasing the use of the train for work-related and leisure displacements must necessarily involve an improvement in the quality of the services, spaces and internal-external paths of the stations. These efforts should also be accompanied by inserting new activities, to be identified according to the profile of each station.

With the aim of rethinking the suburban railway stations as 'stations of every-day life' and places of daily frequentation, services must be designed not only for commuters, but also for the inhabitants and temporary users, in relation to leisure activities, in particular for those stations inside regional parks and areas characterized by historic heritage.

In this framework, the experience carried out by SNCF on the regional railway stations in the Ile de France region is particularly interesting. Some stations, based on their performances and territorial conditions, have been reinterpreted as *tiers-*

lieu and equipped as a space between home-place and work-place, so as to offer differently modulated services for co-working, FabLabs and micro-working spaces (Cerema 2016). The goal, in this case, is to respond to new demands for services and spaces for workers who live the daily displacement as an 'active' moment of their working day, but also to consider the needs of the inhabitants, creating spaces for new activities and stimulating innovative attitudes.

If the results of the classification of the stations allow for decision makers to address and prefigure actions useful for enhancing the use of the train, relevant issues from a governance point of view remain unsolved. The public institutions involved and responsible both in the reorganization of the local public transport and in land-use policies operate at different scales, with different skills and according to weak coordination. It is well known that coordination between public accessibility and land use, thanks to a reorganization of the land use forecasts, instead calls for inter-municipal planning processes, accompanied by effective measures of fiscal equalization. Without them, scattered land use forecasts and soil consumption are combined with a low impact of the railway supply on mobility practices and land use strategies. Nonetheless, the outcomes of the classification of the railway stations offer a preliminary framework for recognizing which interventions and which stations are promising for enhancing the new railway supply, operationalizing the 'land-use transport feedback cycle', also in a perspective able of strengthening the role of regional stations in improving access to urban opportunities and enabling mobility.

References

Amar G, Peny A, Stathopulos N (1991) Formes et fonctions des points de reseaux. Ratp, Paris

Belzer D, Autler G (2002) Transit oriented development: moving from rhetoric to reality. The Brookings Institution Center on Urban and Metropolitan Policy and The Great American Station Foundation, Washington

Bertolini L (1999) Spatial development patterns and public transport: the application of an analytical model in the Netherlands. Plan Pract Res 14(2):199–210. https://doi.org/10.1080/02697459915724

Bertolini L (2012) Integrating mobility and urban development agendas: a ma nifesto. The Planning Review 48(1):16–26. https://doi.org/10.1080/02513625.2012.702956

Calthorpe P (2001) The urban network. Calthorpe Associates, San Francisco. Online available at: https://www.calthorpe.com/sites/default/files/Urban%20Network%20Paper.pdf. Accessed on May 8, 2019

Caset F, Vale DS, Viana CM (2018) Measuring the accessibility of railway stations in the Brussels regional express network: a node-place modelling approach. *Networks and Spatial Economics*, 1–36, https://doi.org/10.1007/s11067-018-9409-y

Center for Transit-Oriented Development (2010 Dec) Performance-based transit-oriented development typology guidebook. Available online http://ctod.org/pdfs/2010PerformanceBasedTODTypologyGuidebook.pdf. Accessed on May 8, 2019

Cerema (2016). *Ateliers partenariaux sur les services dans et autour des gares TER*. Retrieved from http://www.territoires-ville.cerema.fr/ateliers-services-dans-et-autour-des-gares-ter-a1414.html

Cervero R (1995) Satellite new towns: Stockholm's rail-served satellites. Cities 12(1):41–51. https://doi.org/10.1016/0264-2751(95)91864-C

Cervero R (2009) Public transport and sustainable urbanism: global lessons. In: Curtis C, Renne JL, Bertolini L (eds) Transit oriented development: making it happen. Ashgate, Farnham, pp 23–35

Chorus P, Bertolini L (2011) An application of the node place model to explore the spatial development dynamics of station areas in Tokyo. J Transp Land Use 4(1):45–58. https://doi.org/10.5198/jtlu.v4i1.145

Curtis C (2008) Planning for sustainable accessibility: the implementation challenge. Transp Policy 15:104–112. https://doi.org/10.1016/j.tranpol.2007.10.003

Curtis C, Low N (2012) Institutional barriers to sustainable transport. Routledge, London

Curtis C, Renne JL, Bertolini L (eds) (2009) Transit oriented development: making it happen. Ashgate, Aldershot

Dupuy G (1991) L'urbanisme des réseaux: théories et méthodes. Armand Colin, Paris

Evans JE, Pratt RH (2007) Transit oriented development, Transit Cooperative Research Program (TCRP) report 95: Traveler response to transportation system changes handbook, vol 3. Transportation Research Board, Washington

Filion P, McSpurren K (2007) Smart growth and development reality; the difficult coordination of land use and transport objectives. Urban Stud 44(3):501–524. https://doi.org/10.1080/00420980601176055

Haywood R (2005) Co-ordinating urban development, stations and railway services as a component of urban sustainability: an achievable planning goal in Britain? Plan Theory Pract 6(1):71–97. https://doi.org/10.1080/1464935042000334976

Joseph I (ed) (1995) *Gare du Nord. Mode d'emploi.* RATP-Editions Recherches, Paris

Kamruzzaman M, Baker D, Washington S, Turrell G (2014) Advance transit oriented development typology: case study in Brisbane, Australia. J Transp Geogr 34:54–70. https://doi.org/10.1016/j.jtrangeo.2013.11.002

Knowles RD (2006) Transport shaping space: differential collapse in time-space. J Transp Geogr 14(6):407–425

Knowles RD (2012) Transit oriented development in Copenhagen, Denmark: from the finger plan to Ørestad. J Transp Geogr 22:251–261. https://doi.org/10.1016/j.jtrangeo.2012.01.009

Kokoreff M (2002) Pratiques urbaines d'un quartier de gare. Espac Soc 108–109(1):177–196

Lautso K, Wegener M (2007) Integrated strategies for sustainable urban development. In: Marshall S, Banister D (eds) Land use and transport. Elsevier, Amsterdam, pp 153–174

Lyu G, Bertolini L, Pfeffer K (2016) Developing a TOD typology for Beijing metro station areas. J Transp Geogr 55:40–50. https://doi.org/10.1016/j.jtrangeo.2016.07.002

Meyer M, Miller E (2001) Urban transportation planning: a decision-oriented approach. McGraw-Hill, Boston

Newman P (2009) Planning for transit oriented development: strategic principles. In Curtis C, Renne JL, Renne BL (eds) Transit oriented development: making it happen (pp 13–22). Ashgate, Aldershot

Ngo VD (2012) Identifying areas for transit-oriented development in Vancouver using GIS. Trail Six Undergrad J Geogr 6:91–102

Nigro A, Bertolini L, Moccia F (2019) Land use and public transport integration in small cities and towns: assessment methodology and application. J Transp Geogr 74:110–124. https://doi.org/10.1016/j.jtrangeo.2018.11.004

Pojani D, Stead D (2016) A critical deconstruction of the concept of transit oriented development (TOD). In REAL CORP proceedings/Tagungsband, 1–5

Pucci P (2015a) Nouvelle offre ferroviaire dans la Région Urbaine de Milan. In: Grosjean B, Leloutre G, Pucci P, Grillet-Aubert A, Bowie K, Bazaud C (eds) La desserte ferroviaire des territoires périurbains. Éditions Recherches, Paris

Pucci P (2015b) Rincorrere la dispersione: nuova offerta ferroviaria in ambiti a bassa densità insediativa. Il caso delle linee S9 e S13 nella regione urbana Milanese. Territorio 75:117–128

Pucci P (2017) Mobility behaviours in peri-urban areas. The Milan Urban Region case study. Transp Res Proc 25:4233–4248. https://doi.org/10.1016/j.trpro.2017.05.227

Pucci P (2019) Dialogando sui movimenti. La mobilità come capitale spaziale, In: Perrone C, Paba G. (eds) Confini, movimenti, luoghi. Politiche e progetti per città e territori in transizione, Donzelli, Roma

Regione Lombardia (2016 Sep) Programma regionale della mobilità e dei trasporti. Available online http://www.regione.lombardia.it/wps/wcm/connect/775eb7c3-f416-40bc-9449-363670e43a5e/PRMT+inglese.pdf?MOD=AJPERES&CACHEID=775eb7c3-f416-40bc-9449-363670e43a5e. Accessed on May 8, 2019

Renne JL (2009) Measuring the success of transit oriented development. In: Curtis C, Renne JL, Renne BL (eds) Transit oriented development: making it happen (pp 241–255). Ashgate, Aldershot

Renne JL (2009b) From transit-adjacent to transit-oriented development. Local Environ Int J Justice Sustain 14(1):1–15. https://doi.org/10.1080/13549830802522376

Renne JL, Wells JS (2004) Emerging European-style planning in the USA: transit-oriented development. World Transp Policy Pract 10(2):12–24

Reusser DE, Loukopoulos P, Stauffacher M, Scholz RW (2008) Classifying railway stations for sustainable transitions—balancing node and place functions. J Transp Geogr 16(3):191–202. https://doi.org/10.1016/j.jtrangeo.2007.05.004

Rittel H, Webber M (1973 Jun) Dilemmas in a general theory of planning. Policy Sci 4(2):155–169

Singh YJ (2015) Planning for Transit Oriented Development (TOD) using a TOD index. In: 94th annual meeting of the Transportation Research Board, Washington, 11–15 Jan

Singh YJ, Lukmana A, Flackea J, Zuidgeest M, Van Maarseveena MFAM (2017) Measuring TOD around transit nodes—towards TOD policy. Transp Policy 56:96–111. https://doi.org/10.1016/j.tranpol.2017.03.013

Staricco L, Vitale Brovarone E (2018) Promoting TOD through regional planning. A comparative analysis of two European approaches. J Transp Geogr 66:45–52. https://doi.org/10.1016/j.jtrangeo.2017.11.011

Tillous M (2009) Réseaux et territoires vécus. In: Fumey G, Varlet J, Zembri P (eds) Mobilités contemporaines: approches géoculturelles des transports. Ellipses, Paris, pp 43–52

Tillous M (2016) Le métro comme territoire: à l'articulation entre l'espace public et l'espace familier. Flux 103–104(1):32–43

Vale DS (2015) Transit-oriented development, integration of land use and transport, and pedestrian accessibility: combining node-place model with pedestrian shed ratio to evaluate and classify station areas in Lisbon. J Transp Geogr 45:70–80. https://doi.org/10.1016/j.jtrangeo.2015.04.009

Vale DS, Viana CM, Pereira M (2018) The extended node-place model at the local scale: evaluating the integration of land use and transport for Lisbon's subway network. J Transp Geogr 69:282–293. https://doi.org/10.1016/j.jtrangeo.2018.05.004

Van Wee B (2002) Land use and transport: research and policy challenges. J Transp Geogr 10:259–271. https://doi.org/10.1016/s0966-6923(02)00041-8

Wegener M, Fürst F (1999) Land-use transport interaction: state of the art. SSRN Electr J. https://doi.org/10.2139/ssrn.1434678

Zemp S, Stauffacher M, Lang DJ, Scholz RW (2011) Classifying railway stations for strategic transport and land use planning. Context matters! J Transp Geogr 19:670–679. https://doi.org/10.1016/j.jtrangeo.2010.08.008

Chapter 6
The Policy Implications of Enabling Mobilities

Abstract The chapter aims at discussing a suitable policy framework for addressing urban mobility issues from the perspective of enablement, providing operational elements for overcoming the limitations of mainstream transport planning practice and for generating specific collective advantages. Assuming a focus on institutional actors, the framework defines the features that institutions should consider for enhancing enablement through planning and policy approaches, discussing also the role that public subjects should have and the potential benefits deriving from the approach. The proposed framework consists of five elements, discussed more indepth throughout the chapter: the policy aims and ethical principles that may underly planning and policy approaches as well as the design, implementation and evaluation of policy for enabling mobility. Within this scheme, institutions should mobilise the analytical tools necessary to define areas and populations requiring priority interventions and act as facilitators for the deployment of differentiated courses of action. While different aims are pursued when addressing the issues of mobility, a focus on its enabling role is relevant not just for better understanding nor simply for evaluative purposes, but even more so to shape specific policy measures focused on what people use urban mobility for.

Keywords Urban mobility · Transport planning · Urban policy · Policy framework

6.1 Introduction

It is necessary to define how the interest in enabling mobilities can contribute to real-world mobility planning and policy, in order to plan and design transport systems that can contribute to the achievement of individuals' valued purposes. Mainstream transport planning, traditionally based on the prevision of demands and the provision of services, shows limitations in this sense (Martens 2006). Yet the governance condition of a context can also enhance or impede a focus on the social dimensions of

This chapter was authored by Giovanni Vecchio.

© The Author(s), under exclusive license to Springer Nature Switzerland AG 2019
P. Pucci and G. Vecchio, *Enabling Mobilities*, PoliMI SpringerBriefs,
https://doi.org/10.1007/978-3-030-19581-6_6

mobility. The analytical and operational approaches described in the previous chapters must thus be part of a suitable policy framework for addressing urban mobility issues from an enablement perspective. The chapter discusses such a framework, considering the various actors who shape urban mobility but focuses on the role that public institutions can play. These in fact are the subjects traditionally in charge of planning urban mobility and providing infrastructures and services. They are thus the privileged and intended interlocutors for the suggestions proposed in this book.

A policy framework is required to define the features that institutions should consider for enabling mobility. The attempt to bring an ethical concern to real-world operational approaches to mobility can be difficult, require different actors to support it and depends also on the specificities of each setting. The proposed framework consists of five elements, discussed more in depth in the following sections: the policy aims and the ethical principles that may underly planning and policy approaches as well as the design, implementation and evaluation of policy for enabling mobility. The features of the framework are all significant for defining mobility policy, but only some of them can be directly addressed with operational tools. Within this scheme, institutions should mobilise the analytical tools necessary to define areas and populations requiring priority interventions and act as facilitators for the deployment of differentiated courses of action. While different aims are pursued when addressing the issues of mobility, a focus on its enabling role is relevant not just for a better understanding, nor simply for evaluative purposes, but even more so to shape specific policy measures focused on the reasons people use urban mobility. Therefore, the relevance of the approach needs to be supported also by referring to the potential advantages it may provide to institutional actors, allowing them to define priority interventions, shape more precise analyses and tailored operational alternatives as well as better estimate the effects that mobility-related interventions may have on different groups of a population.

6.2   Policy Aims

Aims are the first element that determine policy decisions, according to the purposes that institutions intend to achieve. In relation to mobility, the assumed focus here on enablement defines that the access to valued opportunities should be the main aim to pursue, enhancing also the different features that shape the individual experience of mobility. As previously described, this approach intends to contribute to the individual freedom to live the kind of life that each person has reason to value. However, the valuable opportunities that contribute in this sense may differ and require us to keep plurality in mind. In fact, "the plurality of reasons that a theory of justice has to accommodate relates not only to the diversity of objects of value that the theory recognizes as significant, but also to the type of concerns for which the theory may make room" (Sen 2009, p. 395). In this sense, plurality becomes also a central condition that institutions should guarantee with their action.

Until now, the pursuit of social goals through mobility has received limited yet increasing attention in transport planning and policy. Institutional guidelines for mobility plans explicitly assume a concern for social sustainability, as explicitly stated for example in the European guide for Sustainable Urban Mobility Plans (Arsenio et al. 2016) or in North American plans (Manaugh et al. 2015). Interestingly, planning organisations in the United States seem to adopt quite strong distributional standards, even if the federal guidelines do not provide explicit guidance in this sense (Martens and Golub 2018). Some established planning experiences from Latin America have reshaped complex metropolises intervening on mobility as a tool for social inclusion, as in the renown cases of Curitiba, Bogotá and Medellin (Ardila-Gómez 2004; Brand and Dávila 2011; Vecchio 2017). The same interest in the social dimension of mobility may struggle to find space in traditional participatory processes, as research in Montreal has shown (Boisjoly and Yengoh 2017). However, different cultural traditions can determine a diverse public awareness of social issues, reflected both in policy and public discourse. For example in China, the joint influence of Confucian tradition and socialist ideology gives egalitarianism a central role in the public perception of mobility-related issues, as can be seen in the new congestion charge in Beijing (Liu et al. 2019). Latin American settings may instead be characterised by a tradition of inequality and radical political participation where organised citizens groups bring forward concerns for the social implications of mobility, contesting top-down policies and producing alternative ones, sometimes also through conflict (Sagaris 2014; Sosa López and Montero 2018; Verlinghieri and Venturini 2018). Enabling mobility can thus be the explicit aim of institutional and non-institutional actors intervening in planning and policy processes for mobility, with relevant differences according to the examined setting.

The promotion of enablement, however defined, is nonetheless just one of the aims that transport planning may intend to achieve. While transport planners may be specifically interested in fostering activity participation (Martens 2017b), the different aims achievable with actions concerning mobility are interdependent from each other. Their nature may be mainly political or technical and different actors may be interested in them (for example, politicians, technicians, stakeholders). These aims may be the object or the result of negotiation between the several subjects who are involved or have an interest in mobility-related issues. A focus on enabling mobilities needs to be aware of such manifold aims even if these refer to a pre-planning dimension—that is, the different purposes achievable with transport-related interventions depend on several considerations that only subsequently influence technical decisions.

Other purposes, such as economic growth and environmental sustainability are often at stake when dealing with mobility issues. Transport-related investments in fact can have additional benefits on economic growth at the local and regional level (Banister and Berechman 2001). Their impacts on environmental quality are also relevant, an acknowledged issue that has inspired devoted planning approaches such as the European Sustainable Urban Mobility Plans (Wefering et al. 2014). The relevance of these aims may change according to the examined settings and the actors involved in mobility-related decisions. Furthermore, approaching everyday urban mobility issues as a matter of environmental or economic sustainability frames differently both the definition of the problems to be faced and the consequent operational actions. However, economic and environmental concerns may relate to the social focus on the access to valued opportunities. For example, such access contributes to the individual freedom of choice over alternative lives, an issue that Amartya Sen has extensively discussed as a crucial feature for social development. In his words, development can be defined as freedom (Sen 1999) and individual freedom as a social commitment (Sen 1990).

The social relevance of increased individual freedom suggests that a focus on enablement may provide a more convenient approach to mobility, generating wider advantages that nonetheless must be estimated—at least tentatively—when considering specific planning and policy measures. The evaluation of the social impacts of enablement-based interventions on mobility intercepts also other dimensions. Improvements in accessibility are not simply the result of transport-related investments, but also of the political, policy and institutional conditions that support the definition, implementation and management of such investments (Banister and Berechman 2001). When these features are complemented with positive local economic conditions, growth may emerge because of interventions also in the field of mobility. Moreover, interventions may aim at achieving their purposes by complying with specific requisites. For example, issues of efficiency and effectiveness can have a strong influence on specific mobility-related decisions, particularly in a condition of dominant austerity when interventions that are in line with these criteria can be appealing for decision-makers (Donald et al. 2014).

6.3   Ethical Principles

Assuming enablement as one of the main aims of mobility planning, different ethical principles of reference can differently determine what enablement is. The chosen principles are relevant because these are central for defining the aims of a policy and, consequently, assess to what extent transport systems currently contribute to such aims, define what a desired scenario should be, design the actions to achieve it and finally assess how the costs and benefits generated by such actions are distributed. The academic debate until now has focused on two main concepts of reference: equity and justice. Equity is widely mentioned when considering inequalities and their reflection on transport systems (Van Wee and Geurs 2011). It refers to the distribution of benefits and costs over members of a society and in the case of transport refers to three main elements: what are the benefits and costs, the population groups over which these are distributed and the distributive principles according to which a certain distribution is acceptable or not (Di Ciommo and Shiftan 2017, p. 140). Justice instead is recently gaining space and receives a different definition when considered in relation to transport or mobility. Transport justice in fact refers to the fact that "all members of a society should be guaranteed a sufficient level of accessibility under most, but not all, circumstances" (Martens 2017a, p. 128), using justice as a reference for assessing and planning transport systems. Mobility justice instead undertakes a wider scope, considering mobility as a fundamental condition of being, of space, of subjects and of power (Sheller 2018). In doing so, different forms of justice (not only distributive, but also procedural, deliberative, epistemic and restorative) are considered in relation to different scales—from the individual body to the whole planet. Referring to equity or justice, in relation to transport or mobility more in general, determines a different perspective for considering the social implications of mobilities.

The ethical principles that may shape mobility planning and policy are increasingly debated, from both a theoretical and operational point of view. Nonetheless, "there is still little engagement with theories in political philosophy to frame what justice means in the context of transport policies" (Pereira et al. 2017, p. 170). Such limited engagement is reflected in the growing body of works that use accessibility as a proxy for the way in which transport systems enhance or impede individuals' opportunities. Many works in fact tend to generically consider inequalities, for example considering how accessibility is distributed over different areas of a territory or over different groups of a population. To do so, both established and new measures of inequality are used, such as the Gini coefficient, the Lorenz curves and the Palma ratio (Guzman and Bocarejo 2017; Guzman and Oviedo 2018). Fewer works instead refer explicitly to an ethical approach to generate evaluative frameworks, as was attempted in Chap. 2 of this book. Amongst this body of works, theories such as Rawls' distributive justice, Frankfurt's sufficientarism and Sen's capabilitarianism emerge (Lucas et al. 2016; Pereira 2018, 2019; Tiznado-Aitken et al. 2018).

The explicit statement of what ethical principles are assumed as a reference is central, since different ethical principles determine different distributive outcomes (Martens and Golub 2012). For example, a new public transport line may be differently designed according to the reference equity principle and the target population, as Fernandez Milan and Creutzig (2017) show with four examples:

- *horizontal spatial equity*: adding a new transit connection provides equal access to all areas;
- *vertical economic equity*: the new transit line improves access specifically for the less well-off;
- *vertical gender equity*: the new transit line helps women specifically;
- *vertical social equity*: transit-oriented development improves the social capital of residents.

Different measures may thus have a universalist approach (referring to all the members of a population, irrespectively of their features and differences) or rather promote a 'positive discrimination' (privileging those subjects who are more in need, due to their socioeconomic, gender, physical or age features). A policy framework for enabling mobility should thus be aware that different ethical principles of reference, even implicit ones, have a crucial influence on how we assess existing transport systems and plan for future ones. However, it is significant to consider that different subjects can be responsible for choosing, implicitly or explicitly, the reference ethical principles. These could be practitioners acting according to their technical expertise or in compliance with norms or decision makers such as politicians, whose actions may depend on their political orientation as well as the programs they intend to implement. It could even be citizens at large who may make their preferences explicit by participating in planning processes, promoting devoted initiatives related to mobility issues or voting in devoted consultations.

6.4 The Design of Policy for Enabling Mobilities

While varied aims influence a policy framework for enabling mobilities, planning and policy approaches directly address operational dimensions. We may assume that mobility planning and policy require a process that can be summarised into three elements: design (concerning the definition of the problems to be faced and the actions to address them), implementation (putting such measures into practice) and evaluation (considering outputs and outcomes of the adopted measures). Within such processes, the operational approaches described in the previous chapters play a crucial role.

Analysis is the first aspect that directly involves planning and provides the basis for further interventions. Analytical tools in fact are necessary for assessing how transport systems currently provide each person with the access one needs to valued opportunities, pursuing a basic or equal form of accessibility as a primary aim. According to the activities considered as relevant, as well as to the ethical principles assumed as reference, it is possible to recognize what areas require priority mobility interventions in order to have enough access to opportunities. Assuming a simple and simplified example related to access to jobs in a city, three principles can be referenced—utilitarianism, sufficientarianism and egalitarianism—that may have different aims and, therefore, results:

- *utilitarianism*: maximise overall utility. This approach may privilege the access to better-paid jobs for better-off subjects, since the deriving utility would be higher than providing access to minimum salary jobs to low-income people;
- *sufficientarianism*: provide sufficient access to each area of the city, defining the kind and amount of job opportunities that should be available within a given spatial or temporal distance. This approach would grant that each area has access to at least a minimum amount of job opportunities, to be defined a priori or with a participatory process;
- *egalitarianism*: provide equal access to each area of the city. This approach would grant that each part of the city can access the same number and kind of job opportunities.

The choice of ethical principles also shapes the operational tools presented in the previous chapters. The approach proposed in Chap. 2 draws on aggregate measures of basic accessibility, assuming a sufficientarian approach. Basic accessibility can provide a favourable starting point for institutions, since it is a simple technical tool that can draw on a relatively small set of data to define priorities for action. It is a tool initially based on informed assumptions that can be customized according to specific local needs, considering for example specific valuable opportunities, modal choices, impedance factors and accessibility thresholds. Moreover, measuring potential mobility allows us to consider when insufficient levels of accessibility are the effect of low performances of the existing transport systems, highlighting those areas where mobility planning and policy interventions should focus their attention. This approach, however, is a location-based one and considers where different groups of a population are potentially able to move and what opportunities they may reach. To observe where these groups actually move and what opportunities they can access, the reconstruction of mobility practices can be crucial (Chap. 3), also through the use of new sources of information such as big data (Chap. 4). The representation of potential and actual forms of mobility are complementary and go beyond the traditional transport planning approaches: on the one hand, manifold sources of information provide updated and possibly real-time knowledge of how mobility evolves; on the other hand, the assumption of a clear desired scenario such as the one defined by basic accessibility provides a specific target for policy and does not stick with the simple accommodation of the existing transport demand.

The design of policy can rely on the elements highlighted in the previous sections to define what areas need what forms of intervention. The access to valued opportunities may emerge as a critical issue due to their spatial distribution (a matter of land use planning) and the transport system that should allow access to them (an issue of mobility planning). Insufficient access may be addressed also with long-term actions, providing new infrastructure or intervening on land-use. Assuming a shorter temporal threshold, it may be necessary also to improve the opportunities for mobility that are currently available. In doing so, it can be relevant to intervene on both mobility demand and offer through innovative measures that imply the direct involvement of communities (Vecchio 2018). In the short term, it can be relevant to intervene through territorialisation and individualisation—that is, defining areas that require priority interventions and defining tailored forms of intervention (Bifulco et al. 2008). Therefore, once analytical tools have highlighted areas with insufficient accessibility to a specific set of opportunities, it is possible to design policy measures that improve access.

6.5 The Implementation of Policy for Enabling Mobilities

The implementation of policy for enabling mobility must primarily recognise that manifold actors intervene in the field of mobility, developing varied initiatives to achieve one's own aims. While public institutions are often in charge of providing transport services and infrastructures, together with the normative and planning frameworks for mobility in general, other subjects have a relevant role too. Private companies may manage infrastructures or provide other services, even competing with each other. Organised groups of citizens may be active in relation to specific transport issues, as in the case of commuters' associations and cycling activists.

Additionally, the general public may express its general approval or rejection of existing services or proposed infrastructural projects, representing a powerful force that may support or not the action of public institutional actors. Such multiplicity of actors configures mobility planning as a potential 'trading zone' (Galison 1999, 2013; Geissler et al. 2017), in which shared purposes can bring together subjects with different and sometimes even conflicting aims.

In such a complex scenario, institutions are fundamental for the implementation of policy for enabling mobilities. Their central role in shaping mobility can be here summarised in five points:

- First, institutions may directly develop mobility-related initiatives, for example providing transport services and infrastructures, maintaining transport within the wider realm of public action that characterised the welfare State;
- Second, institutions should set the conditions for other subjects to intervene, working on a normative dimension. Mobility-related initiatives, especially innovative ones that are made possible by unprecedented technology advancements, require devoted regulations and often even challenge existing ones. This is for example the case for ridesharing services, whose existence would be illegal in those settings where taxi services are bonded to the ownership of a license;
- Third, institutional actors may act as facilitators of new initiatives, providing the conditions and eventually the resources that allow their implementation. The coproduction of mobility services serves as an example, where "the process through which inputs used to produce a good or service are contributed by individuals who are not 'in' the same organization" (Ostrom 1996, p. 1073). It requires different forms of capital such as monetary resources to invest (economic capital), skilled people to run services (human capital) and even the trust and sense of community that may inspire and sustain similar initiatives (social capital). While such initiatives are more suitable in areas where local subjects are already active, public institutions may enhance the social and human capital required in such initiatives (Vecchio 2018);
- Fourth, institutions may act as facilitators between other actors in the mobility field who may have conflicting interests. For example, institutions can facilitate the fulfilment of unsatisfied needs, negotiating with a public transport company to serve a badly connected area or manage emerging conflicts, facing the clash between traditional taxi services and emerging ridesharing ones;
- Fifth, institutional subjects should evaluate the costs and benefits of policy measures for enabling mobility. This element is relevant for the adoption and the acceptance of any mobility initiative since the proposed solutions should not be simply technically feasible but must also be recognized as socially useful with institutions supporting their development and diffusion (Feitelson and Samuelson 2004).

Institutional subjects have thus a relevant role for enabling mobility, but they are not neutral nor is their role necessarily central when considering mobility planning and policy. First, it is necessary to consider that the political dimension that characterises all the stages of planning and policy processes implies that institutions are ordinary actors deploying their own tactics and strategies. The actors participating in these processes and the interactions they establish between them are thus crucial for the shaping of policy approaches focused on enabling mobilities. Moreover, institutional subjects are manifold and raise governance issues in cooperation between different bodies, which may struggle between themselves or may also require structural changes to better cope with emerging mobility issues (as in the case of metropolitan authorities for mobility, whose field of action often goes beyond the borders of traditional administrative bodies). Other elements, such as the scarcity of public resources and the emergence of new actors, limit the possibility of action for institutional subjects. Under these conditions, interaction with the mobility-related initiatives developed by other subjects becomes even more relevant. This is for example the case for information-related initiatives, which require specific attention to their policy dimension. These initiatives can be relevant also for addressing urban mobility issues but yet require devoted forms of interaction with powerful actors such as corporations (Vecchio and Tricarico 2019). The multiplicity of actors in the mobility policy arena shows that institutional subjects cannot realistically be the only subject in charge of dealing with such a policy field. Rather, forms of 'collaborative governance' involving manifold actors and their initiatives become increasingly relevant for dealing with mobility issues.

6.6 The Evaluation of Policy for Enabling Mobilities

While a focus on enabling mobilities provides a different, socially-relevant focus on planning and policy for mobility, the related operational measures generate benefits and costs, though the evaluation is not straightforward. Assuming the provision of sufficient accessibility as the main aim of urban mobility planning and policy, this would be the target that a measure should be able to achieve. The benefits of a measure would thus depend on the number of people who would be brought above the current sufficiency threshold for accessibility, eventually assuming that the generated benefit is inversely proportional to the current socioeconomic condition of the person. This approach is in line with proposed 'justice tests' that consider the correlation between changes in accessibility and the prevailing socioeconomic outcomes (Nazari Adli and Donovan 2018). The generated benefits may be tentatively quantified attaching a monetary value to saved travel time or to each additional new opportunity that can be gained, even if this monetary-based approach is prone to limitations and should also consider the diminishing marginal value of increased accessibility (Martens 2006). In case a varied set of feasible measures could bring the targeted area or population above the sufficiency threshold, the evaluation would take into account their costs. For example, if both a new bus line and a new metro line were able to improve the basic accessibility available to one neighbourhood, most likely the former, cheaper option would be preferable.

The definition of priority areas of intervention also allows a rough estimation of the benefits that improved accessibility may generate. For example, it is possible to estimate how many people would experience sufficient accessibility thanks to specific interventions. The individual socioeconomic conditions are crucial to assess which subjects rely on public transport for their mobility needs and, consequently, who is going to actually benefit from a certain intervention. This can be seen when high-income subjects compensate an eventual insufficient accessibility by using private vehicles, while the same alternative would be challenging for low-income people. Instead, it is more difficult to estimate the actual benefit generated when achieving a sufficient level of accessibility. For example, it may be assumed that an overall improvement is generated when achieving sufficient access to a given set of opportunities. The estimation may also attach a monetary value both to the generated travel time gains and to each new reachable opportunity. This method is nonetheless prone to several limitations and simplifications, especially in reducing potential material and immaterial gains to a simple monetary dimension. While the possibility to reach a job has not only monetary outcomes (e.g. receiving a salary), it also has significant immaterial consequences (such as the dignity associated with having a job). These considerations assume an improvement in basic accessibility, but they do not yet consider what specific intervention would be defined to achieve it.

However, the emphasis on the evaluation of economic benefits and costs may be problematic from the very perspective of enabling mobilities. The approach proposed in the book emphasises the importance of granting accessibility as the main aim of planning and policy. Due to its social relevance, accessibility may be the object of decommodification, that is, to be considered "as a social entitlement rather than as a product with monetary value that depends solely on market fluctuations" (Hernandez 2017, p. 152). In this sense, public transport is the main 'decommodifier' of accessibility, and its contribution in this sense depends on the coverage of a transport system, its affordability and the eventual existence of devoted mechanisms of financial decommodification. These could include subsidies to lower the general fare, benefits for low-income groups or diversified fares according to space and time (Hernandez 2017, pp. 164–166). Though such actions would generate a social benefit (in terms of increased accessibility), they may also result in higher costs (in terms of increased economic expense for public transport). The evaluation of policy for enabling mobilities therefore cannot exclusively depend on economic features, but must also consider social elements, requiring explicit political decisions in this sense.

6.7 Conclusions

The policy framework has tried to consider different elements that concur when designing, implementing and evaluating policy for enabling mobilities. Institutions should define as well as create consensus for the policy aims and the principles they pursue, their mobilisation of the analytical tools to define areas and populations requiring priority interventions and in acting as facilitators for the deployment of differentiated courses of action. Once these measures have been put into practice, it is necessary to evaluate them, in a process that periodically considers how existing transport systems support individual capabilities providing access to valued opportunities. While the existing systems should be continuously improved by the adoption of new measures, their performance may change if new evaluative criteria are adopted.

The proposed approach seems to configure several advantages for institutions intervening on urban mobility, helping to define priority action and in considering a wider range of operational options to deploy that go beyond the simple construction of new infrastructures or provision of new services. However, planning and policy measures for enabling mobilities are prone to limitations. These measures are not sufficient for addressing individual and collective mobility needs, especially when considering their operational implications. Other elements that influence the functioning of urban mobility require traditional approaches. This is the case for the spatial distribution of significant activities to be reached, or for infrastructures addressing huge mobility flows. Other structural conditions may prove problematic, such as difficult spatial conditions (for example orography) and a lack of adequate infrastructures (like roads) that present obstacles to the provision of ordinary services. Also, the policy arena of a setting may be an obstacle, especially when the actors lack a tradition of cooperation or at least meaningful interactions between each other.

In conclusion, a focus on enabling mobilities appears as suitable for addressing contemporary mobility issues. Its contribution refers to manifold dimensions, from being a suitable ethical orientation for policy measures intending to increase individuals' opportunities to shaping significant operational options. This approach requires grounding such generic reflections in the various settings where they could be significant as well as considering the local needs and specificities that may make such focus more relevant. Nonetheless, the main element of interest seems to be the possibility to consider enabling mobilities not just to better understand existing transport systems and mobility practices nor simply for evaluative purposes but rather to shape specific policy measures focused on what people use urban mobility for.

References

Ardila-Gómez A (2004) Transit planning in Curitiba and Bogotá. Roles in interaction, risk, and change. PhD Thesis, Massachusetts Institute of Technology

Arsenio E, Di Martens K, Ciommo F (2016) Sustainable urban mobility plans: bridging climate change and equity targets? Res Transp Econ 55:30–39. https://doi.org/10.1016/j.retrec.2016.04.008

Banister D, Berechman Y (2001) Transport investment and the promotion of economic growth. J Transp Geogr 9(3):209–218. https://doi.org/10.1016/S0966-6923(01)00013-8

Bifulco L, Bricocoli M, Monteleone R (2008) Activation and local welfare in Italy: trends and issues. Soc Policy Adm 42(2):143–159. https://doi.org/10.1111/j.1467-9515.2008.00600.x

Boisjoly G, Yengoh GT (2017) Opening the door to social equity: local and participatory approaches to transportation planning in Montreal. Eur Transp Res Rev 9(3):43. https://doi.org/10.1007/s12544-017-0258-4

Brand P, Dávila JD (2011) Mobility innovation at the urban margins. City 15(6):647–661. https://doi.org/10.1080/13604813.2011.609007

Di Ciommo F, Shiftan Y (2017) Transport equity analysis. Transp Rev 37(2):139–151. https://doi.org/10.1080/01441647.2017.1278647

Donald B, Glasmeier A, Gray M, Lobao L (2014) Austerity in the city: economic crisis and urban service decline? Camb J Reg Econ Soc 7(1):3–15. https://doi.org/10.1093/cjres/rst040

Feitelson E, Samuelson I (2004) The political economy of transport innovations. In: Beuthe M, Himanen V, Reggiani A, Zamparini L (eds) Transport developments and innovations in an evolving world. Springer, Berlin, pp 11–26

Fernandez Milan BF, Creutzig F (2017) Lifting peripheral fortunes: upgrading transit improves spatial, income and gender equity in Medellin. Cities 70:122–134. https://doi.org/10.1016/j.cities.2017.07.019

Galison P (1999) Trading zone: coordinating action and belief. In: Biagioli M (ed) The science studies reader. Routledge, London, pp 137–160

Galison P (2013) Trading plans. In: Mäntysalo R, Balducci A (eds) Urban planning as a trading zone. Springer, Dordrecht, pp 195–207

Geissler J-B, Tricarico L, Vecchio G (2017) The construction of a trading zone as political strategy: a review of London infrastructure plan 2050. Eur J Spat Dev 64:1–22

Guzman LA, Bocarejo JP (2017) Urban form and spatial urban equity in Bogota, Colombia. Transp Res Proc 25:4491–4506. https://doi.org/10.1016/j.trpro.2017.05.345

Guzman LA, Oviedo D (2018) Accessibility, affordability and equity: assessing 'pro-poor' public transport subsidies in Bogotá. Transp Policy 68:37–51. https://doi.org/10.1016/j.tranpol.2018.04.012

Hernandez D (2017) Public transport, well-being and inequality: coverage and affordability in the city of Montevideo. CEPAL Rev 122:152–169

Liu Q, Lucas K, Marsden G, Liu Y (2019) Egalitarianism and public perception of social inequities: a case study of Beijing congestion charge. Transp Policy 74:47–62. https://doi.org/10.1016/j.tranpol.2018.11.012

Lucas K, Van Wee B, Maat K (2016) A method to evaluate equitable accessibility: combining ethical theories and accessibility-based approaches. Transportation 43(3):473–490. https://doi.org/10.1007/s11116-015-9585-2

Manaugh K, Badami MG, Elgeneidy AM (2015) Integrating social equity into urban transportation planning: a critical evaluation of equity objectives and measures in transportation plans in North America. Transp Policy 37:167–176. https://doi.org/10.1016/j.tranpol.2014.09.013

Martens K (2006) Basing transport planning on principles of social justice. Berkeley Plann J 19:1–17

Martens K (2017a) Transport justice: designing fair transportation systems. Routledge, New York, London

Martens K (2017b) Why accessibility measurement is not merely an option, but an absolute necessity. In: Punto N, Hull A (eds) Accessibility tools and their applications. Routledge, New York, London

Martens K, Golub A (2012) A justice-theoretic exploration of accessibility measures. In: Geurs KT, Krizek KJ, Reggiani A (eds) Accessibility analysis and transport planning: challenges for Europe and North America. Elgar, Celtenham, pp 195–210

Martens K, Golub A (2018) A fair distribution of accessibility: interpreting civil rights regulations for regional transportation plans. J Plann Educ Res. https://doi.org/10.1177/0739456x18791014

Nazari Adli S, Donovan S (2018) Right to the city: applying justice tests to public transport investments. Transp Policy 66:56–65. https://doi.org/10.1016/j.tranpol.2018.03.005

Ostrom E (1996) Crossing the great divide: coproduction, synergy, and development. World Dev 24(6):1073–1087. https://doi.org/10.1016/0305-750x(96)00023-x

Pereira RHM (2018) Transport legacy of mega-events and the redistribution of accessibility to urban destinations. Cities 81:45–60. https://doi.org/10.1016/j.cities.2018.03.013

Pereira RHM (2019) Future accessibility impacts of transport policy scenarios: equity and sensitivity to travel time thresholds for bus rapid transit expansion in Rio de Janeiro. J Transp Geogr 74:321–332. https://doi.org/10.1016/j.jtrangeo.2018.12.005

Pereira RHM, Schwanen T, Banister D (2017) Distributive justice and equity in transportation. Transp Rev 37(2):170–191. https://doi.org/10.1080/01441647.2016.1257660

Sagaris L (2014) Citizen participation for sustainable transport: the case of "Living City" in Santiago, Chile (1997–2012). J Transp Geogr 41:74–83. https://doi.org/10.1016/j.jtrangeo.2014.08.011

Sen AK (1990) Individual freedom as social commitment. India Int Cent Q 17(1):101–115

Sen AK (1999) Development as freedom. Oxford University Press, Oxford

Sen AK (2009) The idea of justice. Harvard University Press, Cambridge

Sheller M (2018) Mobility justice: the politics of movement in an age of extremes. Verso, London

Sosa López O, Montero S (2018) Expert-citizens: producing and contesting sustainable mobility policy in Mexican cities. J Transp Geogr 67:137–144. https://doi.org/10.1016/j.jtrangeo.2017.08.018

Tiznado-Aitken I, Muñoz JC, Hurtubia R (2018) The role of accessibility to public transport and quality of walking environment on urban equity: the case of Santiago de Chile. Transp Res Rec 2672(35):129–138. https://doi.org/10.1177/0361198118782036

Van Wee B, Geurs K (2011) Discussing equity and social exclusion in accessibility evaluations. Eur J Transp Infrastruct Res 11(4):350–367

Vecchio G (2017) Democracy on the move? Bogotá's urban transport strategies and the access to the city. City Territory Architect 4(15):1–15. https://doi.org/10.1186/s40410-017-0071-3

Vecchio G (2018) Producing opportunities together: sharing-based policy approaches for marginal mobilities in Bogotá. Urban Sci 2(3):54. https://doi.org/10.3390/urbansci2030054

Vecchio G, Tricarico L (2019) "May the force move you": roles and actors of information sharing devices in urban mobility. Cities 88:261–268. https://doi.org/10.1016/j.cities.2018.11.007

Verlinghieri E, Venturini F (2018) Exploring the right to mobility through the 2013 mobilizations in Rio de Janeiro. J Transp Geogr 67:126–136. https://doi.org/10.1016/j.jtrangeo.2017.09.008

Wefering F, Rupprecht S, Bührmann S, Böhler-Baedeker S (2014) Guidelines. Developing and implementing a sustainable urban mobility plan. Brussels

Chapter 7
Open Directions for Enabling Mobilities

Abstract The chapter discusses further directions for research on the social implications of transport planning decisions, considering three dimensions that can enhance existing analytical and operational tools for enabling mobilities. First, the guiding concepts orienting evaluative and operational approaches require a specific definition in light of the contribution mobility can give to individuals and their opportunities. Second, existing tools need to be reshaped to emphasize the social implications of mobility. Third, the real-world conditions that influence policymaking processes in the domain of urban and transport issues deserve greater attention, together with the specificities of any given territorial setting. The theoretical relevance of socially-oriented approaches to mobility has been widely discussed, together with the potential operational inflections that they may receive. However, these three dimensions for future research may instead contribute to making such approaches feasible, replicable and able to influence the consolidated approaches to transport planning and policy.

Keywords Transport planning · Transport policy · Enablement

7.1 Introduction

The research experiences, reported in the previous chapters, share the same interest in mobility as a potential enabling condition, which can impact the opportunities available to each person. The discussed approaches deal with different analytical and planning tools, addressing heterogeneous issues and opportunities in diverse settings.

The intended multiplicity of the chapters somehow reflects that mobility planning and policy are increasingly assuming an interest in the enabling role of mobility and developing it in different directions, moving away from their traditional approach based on the prevision of demands and the provision of services. The proposed focus on the enabling role of mobility contributes to addressing the issues of everyday

---

This chapter was co-authored by Giovanni Vecchio and Paola Pucci.

© The Author(s), under exclusive license to Springer Nature Switzerland AG 2019
P. Pucci and G. Vecchio, *Enabling Mobilities*, PoliMI SpringerBriefs,
https://doi.org/10.1007/978-3-030-19581-6_7

mobility by fostering a systematic understanding of its social relevance and providing operational approaches that can take these aspects into account.

These approaches can provide an interpretative advancement in respect to the current generalized difficulty in interpreting the urban in general (Brenner and Schmid 2015) and mobility in particular, due to its socio-spatio-temporal dimensions (Dodier 2013; Kaufmann 2002; Pucci 2016) and the increasing multiplicity of related needs and practices (Adey and Bissel 2010; Cresswell 2010; Hannam et al. 2006). These conclusive remarks intend thus to stress the significance of an approach to urban mobility focused on its enabling role by proposing some open directions for research in terms of orienting evaluative and operational tools for addressing policymaking processes in the domain of urban and transport issues. To do so, three dimensions are considered. First, the guiding concepts orienting evaluative and operational approaches require a shared specific definition in light of the contribution mobility can give to individuals and their opportunities. Second, existing tools need to be reshaped according to the opportunity to deal with the social implications of mobility. Third, the real-world conditions that influence policymaking processes in the domain of urban and transport issues deserve greater attention, together with the specificities of any given territorial setting. The theoretical relevance of socially-oriented approaches to mobility has been widely discussed, together with the potential operational inflections that they may receive. These three dimensions for future research may instead contribute to making such approaches feasible, replicable and able to influence the consolidated approaches to transport planning and policy.

7.2   Conceptual Issues: What Do Enabling Mobilities Mean?

Approaching mobility issues from the perspective of enablement can provide significant improvements to the knowledge used by planning and policy approaches. A crucial element is the focus on what mobility is and could be used for. As recalled in this book, mobility is not simply a phenomenon that concludes in itself, based on overcoming the spatial friction between point A and point B. Rather, it can be seen as a specific ability any person may form and use to gain wider opportunities. In addition, because mobility has never been synonymous with freedom and citizenship (Cresswell 2004, p. 145), a more complex understanding of people in relation to mobility and to transport systems is required. Traditional transport planning approaches in fact tend to consider subjects and in some cases even their individual features, but mainly according to the travel behaviours they may develop (Dijst et al. 2013) and their consequent aggregate reflections on transport volumes (Van Wee 2013). More recently, the diffusion of information technologies supports the idea of 'quantified traveller' (Jariyasunant et al. 2015), whose mobility can be grasped and even changed by 'the digital skin of cities' (Rabari and Storper 2015, p. 28). These approaches consider people only for their present reflection on actual cities. Transport volumes or big data consider those subjects who are already able to move, reach a place and take part in an activity, ignoring instead those that are not able or willing to move for a variety of reasons.

The focus on enabling mobilities requires us nonetheless to consider that at least three conceptual and interpretative limitations affect its potential contribution to planning and policy due to the heterogeneity of reference concepts, their difficult operationalisation and the tension between individual needs and universalistic approaches to them. Different streams of the existing literature on the social implications of mobility tend to develop heterogeneous reference frameworks. As recalled in Chaps. 1 and 6, there are diverse definitions given to the issue to be addressed (social exclusion, inequality, injustice), the reference concepts to adopt (equity, equality, justice) and the role that mobility can play (as a resource, an ability or as a good valuable in itself). These approaches tend to overlook "the broader implication of a comprehensive transport policy" (Beyazit 2011, p. 130), calling for a deeper exploration of their operational implications—both in terms of analysis and policy design—and the necessary adaptations to translate these tools into planning and policy practice. Moreover, a focus on enabling mobilities implies a highly subjective dimension as what each person values differs according to the subject considered and is often difficult to grasp using the prevailing evaluative aggregate approaches (Comim 2008). Finally, a significant procedural problem emerges when translating into reality the interest for enabling mobilities, as mobilities are "striated by a whole series of rules, conventions and institutions of regulation and control" (Amin and Thrift 2002, p. 26).

Technicians, politicians or the general public may give different definitions as to what constitutes a fair transport system. Also, different methods could be used for making decisions in this sense from relying simply on informed assumptions by experts, following the political will of the decision makers currently in charge or establishing devoted participatory processes involving citizens. Therefore, future research should strongly engage also with real world processes, seeing in practice what such concepts mean in different settings and for different actors. The procedural dimension of transport planning and policy can thus be crucial for defining and refining what enabling mobilities mean.

7.3 Operational Issues: Planning for Enabling Mobilities

A focus on enabling mobilities provides firstly an advancement of the knowledge available to us by involving features that are often overlooked and by using them to readdress existing evaluative tools and combining different evaluative approaches. While the studies found in this book deal with established tools (from accessibility evaluations, the reconstruction of mobility practices, planning railway stations to the design of urban policy), a focus on enablement can enrich such tools and combine them. The approaches presented in the book provide the basis for an alternative operational toolkit, whose use depends on the wider policy framework they are part of.

From the perspective of enablement, mobility is approached having in mind that "movement makes connections and connections make inequalities" (Urry 2012, p. 24). The aims to which the toolkit concur are therefore the result of a specific problem framing (Schön 1983) or how a situation is set and consequently addressed. The focus assumed in this book privileges a socio-spatial-temporal analysis of mobilities, even if other societal goals may concur or compete (for example, the pursuit of environmental sustainability or economic growth through the management of urban transport systems).

The manifold tools presented in the previous chapters test possible operational dimensions of enabling mobilities and, despite not proposing a straightforward operational protocol, in action they present different applications aimed at offering open and adjustable 'descriptions for' addressing mobility issues.

The resulting findings suggest an incremental strategy for enabling mobility that can facilitate the most finalized problem setting through a purposeful use of simple tools, such as simplified accessibility evaluations, analyses of available data including origin/destination surveys or assessments of existing infrastructures and equipment (Chaps. 2, 3 and 5 provide examples in this sense).

Enabling mobilities is a suitable reference for orientating the evaluation of existing situations and the design of necessary actions. If easily translated into the mainstream transport planning and policy practice, the issue of adaptability must be raised. In this direction, three critical aspects emerge. First, there is the interaction of the proposed tools with established approaches to transport issues, such as the 4-step models or cost-benefit analyses. The solidity and relatively easy use of the mainstream transport planning practice makes it refractory to the social dimensions of mobility, which struggle to be included within these analytical and appraisal tools. Second, the focus on enablement is sometimes difficult to communicate to decision makers and the general public. Third, the lack of data is sometimes a key barrier, referring to both quantitative and qualitative forms of knowledge that can be relevant for understanding how transport systems currently contribute to individuals' opportunities. Future research should consider how analytical and operational tools can assume a simple form, easily adoptable within established transport planning and policy approaches that intend to promote enablement. If more relevant and feasible approaches can be adopted, they could mobilise refined quantitative and qualitative analytical tools (from big data to ethnographic approaches). In this perspective, the interaction with practitioners and policymakers is relevant to better refine such tools according to the needs and the habits of their potential users.

7.4 Policy and Political Issues: Enabling Mobilities in Real-World Processes

The approach outlined until now intends to better understand and address urban mobility issues, but has not specified who should be responsible for doing so. The policy framework outlined in Chap. 6 addresses particularly public institutions, the subjects traditionally in charge of planning transport systems, as well as (in many cases, at least) the provision of infrastructures, running services and the responsibility for maintenance tasks. Nonetheless, the discussed analytical and operational tools involve a range of actors that goes well beyond institutional subjects. While varied may be the actors in the urban mobility policy arena, their actual involvement in operational measures may depend on the contribution they provide. These subjects potentially could mobilise an unprecedented quantity of material and immaterial resources. Therefore, going beyond the simple call for participatory planning approaches and their limited focus on procedural issues seems necessary to configure a direct involvement of these new subjects, not just in raising issues or designing solutions, but also in their very implementation.

The multiplicity of potentially relevant actors also suggests that the features of real-world policymaking are significant for enabling mobility. The governance dimension of planning and policy is often overlooked when considering transport issues (Marsden and Reardon 2017), but it is crucial to understand how current processes operate and how to govern future transitions (Docherty et al. 2017). The social relevance of research on enabling mobilities must therefore consider how to enhance the policy usability of technical tools, as the previous chapters attempted.

In doing so, a first element of attention is the way in which we understand mobility policy. The concept of policy cycle, for example, can easily involve the various steps through which a public problem is faced—from its definition to the evaluation of the achieved result, but risks promoting a linear and unrealistic view of processes that more often go through different cycles of learning and re-discussion of previous assumptions.

This issue refers to the experimental dimension that policy can have (Concilio et al. 2019; Pucci et al. 2019) and, at the same time, raises a second element of attention, referred to as the political dimension of urban experimentations (Savini and Bertolini 2019). In fact, these steps do not simply test new approaches to emerging issues, but rather face innovative experiences on the base of diverse political biases and normative assumptions, which may result in a positive or negative attitude towards innovations. Aiming at enabling mobility, the technical and the political dimensions of planning and policy need to receive the same attention.

7.5 Conclusions

The studies included in this book have attempted to explore the relationship between mobility and enablement, in order to improve how transport planning and policy can contribute to increasing the opportunities available to each one of us. Together with the theoretical and practical issues that deserve further research, a pressing question underlies this final section: how to assure that institutions, together with other relevant actors, are interested in enabling mobilities and act consequently, also when dealing with transport planning and policy? Any new knowledge that intends to have whatever small impact on the real world needs to be at least aware of this dimension. In this sense, it is necessary to demonstrate that it is feasible and relevant to address urban mobility while considering enablement. As for feasibility, this work has hopefully provided some useful elements that work towards this direction. As for relevance, many elements may contribute to supporting it: theoretical debates, empirical data, even first-hand experiences. All these elements once more reflect the richness of each person and, when combined, the complexity of contemporary urban life.

References

Adey P, Bissel D (2010) Mobilities, meetings, and futures: an interview with John Urry. Environ Plann D Soc Space 28:1–16. https://doi.org/10.1068/d3709

Amin A, Thrift N (2002) Cities. Reimagining the urban. Polity, Cambridge

Beyazit E (2011) Evaluating social justice in transport: lessons to be learned from the capability approach. Transp Rev 31(1):117–134. https://doi.org/10.1080/01441647.2010.504900

Brenner N, Schmid C (2015) The epistemology of urban morphology. City 19(2–3):151–182. https://doi.org/10.1080/13604813.2015.1014712

Comim F (2008) Measuring capabilities. In: Comim F, Qizilbash M, Alkire S (eds) The capability approach: concepts, measures and applications. Cambridge University Press, Cambridge, pp 157–200. https://doi.org/10.1017/cbo9780511492587.007

Concilio G, Pucci P, Vecchio G, Lanza G (2019) Big data and policymaking: between real time management and the experimental dimension of policies. In: Lecture Notes in Computer Science LNCS, ICCSA Proceeding, Saint Petersburg

Cresswell T (2004) Justice sociale et droit à la mobilité. In: Allemand S, Ascher F, Lévy J (eds) Les sens du mouvement. Belin, Paris, pp 145–153

Cresswell T (2010) Towards a politics of mobility. Environ Plann D Soc Space 28(1):17–31. https://doi.org/10.1068/d11407

Dijst M, Rietveld P, Steg L (2013) Individual needs, opportunities and travel behaviour: a multidisciplinary perspective based on psychology, economics and geography. In: Van Wee B, Annema JA, Banister D (eds) The transport system and transport policy. Elgar, Cheltenham, pp 19–50

Docherty I, Marsden G, Anable J (2017) The governance of smart mobility. Transp Res Part A Policy Pract 115:114–125. https://doi.org/10.1016/j.tra.2017.09.012

Dodier R (2013) Modes d'habiter périurbains et intégration sociale et urbaine. EspacesTemps.net. Retrieved 10 Mar 2017, from http://www.espacestemps.net/articles/modes-dhabiter-periurbains-et-integration/

Hannam K, Sheller M, Urry J (2006) Editorial: mobilities, immobilities and moorings. Mobilities 1(1):1–22. https://doi.org/10.1080/17450100500489189

Jariyasunant J, Abou-Zeid M, Carrel A, Ekambaram V, Gaker D, Sengupta R, Walker JL (2015) Quantified traveler: travel feedback meets the cloud to change behavior. J Intell Transp Syst 19(2):109–124. https://doi.org/10.1080/15472450.2013.856714

Kaufmann V (2002) Re-thinking mobility. Ashgate, Farnham

Marsden G, Reardon L (2017) Questions of governance: rethinking the study of transportation policy. Transp Res Part A Policy Pract 101:238–251. https://doi.org/10.1016/j.tra.2017.05.008

Pucci P (2016) Mobility practices as a knowledge and design tool for urban policy. In: Pucci P, Colleoni M (eds) Understanding mobilities for designing contemporary cities. Springer, Berlin, pp 3–22

Pucci P, Vecchio G, Concilio G (2019). Big data and urban mobility: a policy making perspective. In: Transportation Research Procedia, WCTR 2019, Mumbai

Rabari C, Storper M (2015) The digital skin of cities: urban theory and research in the age of the sensored and metered city, ubiquitous computing and big data. Cambridge J Reg Econ Soc 8(1):27–42. https://doi.org/10.1093/cjres/rsu021

Savini F, Bertolini L (2019) Urban experimentation as a politics of niches. Environ Plann A Econ Space. https://doi.org/10.1177/0308518x19826085

Schön D (1983) The reflective practitioner. Basic Books, New York

Urry J (2012) Social networks, mobile lives and social inequalities. J Transp Geogr 21:24–30. https://doi.org/10.1016/j.jtrangeo.2011.10.003

Van Wee B (2013) The traffic and transport system and effects on accessibility, the environment and safety: an introduction. In: Van Wee B, Annema JA, Banister D (eds) The transport system and transport policy. Elgar, Cheltenham, pp 4–18

Printed in the United States
By Bookmasters